MICRORNA PROFILING IN CANCER
A Bioinformatics Perspective

MicroRNA Profiling in Cancer

A Bioinformatics Perspective

Edited by Yuriy Gusev

Published by

Pan Stanford Publishing Pte. Ltd.
Penthouse Level, Suntec Tower 3
8 Temasek Boulevard
Singapore 038988

Email: editorial@panstanford.com
Web: www.panstanford.com

British Library Cataloguing-in-Publication Data
A catalogue record for this book is available from the British Library.

MICRORNA PROFILING IN CANCER
A Bioinformatics Perspective

Copyright © 2010 by Pan Stanford Publishing Pte. Ltd.

All rights reserved. This book, or parts thereof, may not be reproduced in any form or by any means, electronic or mechanical, including photocopying, recording or any information storage and retrieval system now known or to be invented, without written permission from the Publisher.

For photocopying of material in this volume, please pay a copying fee through the Copyright Clearance Center, Inc., 222 Rosewood Drive, Danvers, MA 01923, USA. In this case permission to photocopy is not required from the publisher.

ISBN-13 978-981-4267-01-4
ISBN-10 981-4267-01-5

Printed in Singapore by Mainland Press Pte Ltd.

To the Memory of my Mentor
Dr. Andrei Yakovlev

PREFACE

There is a new kid on the block taking the field of cancer research by storm: a new type of noncoding short single stranded RNA molecule, microRNA, has recently emerged as one of the most important new classes of cellular regulators. First discovered in 1993 by Victor Ambros in *C. Elegans*, these small RNAs were termed microRNAs in 2001. Just seven years later the importance of microRNA discovery has been highlighted by the Lasker Award given to Victor Ambros, Gary Ruvkun, and David Baulcombe in 2008.

It is already evident that the discovery of microRNA has created a paradigm shift in post-genomics biology, not only for scientists accustomed to traditional central dogma of molecular biology but also for researchers studying human diseases and accustomed to traditional genetics approach of studying one gene at the time. The ability of microRNAs to control large groups of genes and impose global post-transcriptional regulation of many (if not all) important cellular processes in development, cell proliferation and differentiation has opened up a new dimension and uncovered an astonishing complexity of intracellular regulatory circuitry.

With the advancement of high throughput technologies over past 3 years it is now possible to profile and sequence the entire human microRNome. Latest advances in microRNA research have provided strong evidence that aberrant expression of specific miRNAs is associated with a broad spectrum of human diseases, such as cancer, diabetes, cardiovascular diseases, psychological disorders, and others. First evidence that microRNA could be involved in cancer came in 2002. The number of publications devoted to microRNA in cancer has been growing exponentially ever since. In 2008 alone there were nearly 600 original research papers published on the role of miRNA in cancer with over 160 papers related to miRNA profiling in cancer.

The global microRNome profiling has shown drastic changes in expression of multiple miRNAs in all human cancers tested to date. microRNA expression signatures have often provided a more accurate method of classifying cancer subtypes than transcriptome profiling of all known protein-coding genes. microRNA profiling also allows classification of different stages in tumor progression and in some cases can predict outcome of a disease. As researchers continue to study microRNA expression in cancer and focus on the most relevant microRNAs, the further advancement of microRNA detection technologies and computational bioinformatics methodologies is critical for successful identification of most informative diagnostics markers and drug targets among the tissue specific microRNAs.

This book presents a collection of invited chapters written by experts in the emerging field of miRNomics from the unique prospective of bioinformatics, computational and systems biology. The field is still in flux and our knowledge and understanding of microRNAs is continuously and rapidly evolving. The idea behind this project was to assemble a snapshot of current state of the art microRNA research in a form of critical reviews and summaries of recently published results on microRNA profiling in cancer.

This volume covers a wide range of topics related to microRNA profiling which are grouped in four overlapping categories:

i. Chapters 1 through 4 provide detailed technical guides as well as tricks and tips for using three major technology platforms that are currently available for microRNA profiling: Detection of microRNA with Agilent microarrays (chapter 1) and Exiqon LNA-enhanced microarrays (chapter 2); High throughput real-time quantitative PCR (chapter 3); and Next-Generation sequencing (chapter 4)

ii. Chapters 4, 5, 7 and 9 provide overviews and critical analysis of microRNA expression for several human cancers such as hematological malignancies (chapter 4, 7), prostate cancer (chapter 5), colon, lung, breast, pancreatic and liver cancers (chapter 9).

iii. Chapters 6 and 7 present in depth reviews of microRNome architecture discussing the function of intronic microRNAs and intriguing hypothesis of "a kiss of microRNA" (chapter 6) as well as genome-wide analysis of microRNA genes location showing striking correlation with genomically unstable loci and retroviral integration sites (chapter 7).

iv. Final section of the book is devoted to the advanced systems biology topics on analysis of microRNA regulatory networks and their interconnection with the transcription factors-mediated gene networks (chapter 8), functional profiling of co-expressed microRNAs in cancer and pathway analysis of microRNA targets (chapter 9), as well as important progress in mathematical modeling and computational analysis of dynamics of post-transcriptional gene regulation by microRNAs (chapter 10).

To summarize, this first-of-its-kind book provides a comprehensive review of well over a hundred recent publications in the area of microRNA profiling in cancer with emphasis on system-wide analysis of microRNA data. The chapters are accompanied with nearly 50 well thought out illustrations providing clear and easy-to-understand visualization of complex and novel concepts. The Index and Glossary sections make it easier to navigate through the book.

This book is designed for a broad spectrum of readership of academic, biotech and pharmaceutical industry scientists, physicians involved in cancer research, oncologists, computational biologists, and bioinformaticians who are involved in basic, translational and clinical research, as well as college educators, undergraduate, graduate and medical students.

It could serve not only as an excellent technical guide for principal investigators who are interested in utilizing microRNA profiling in their research, but also as a reference book for college educators wishing to incorporate microRNA-related topics in their computational biology and bioinformatics classes.

I would like to express my sincerest gratitude to the invited authors who made this book possible — for their enthusiastic support of this project, and to many authors on whose work we have drawn in putting together this review volume. I am deeply indebted to my colleagues — professor Daniel Brackett and Megan Lerner, for their unwavering support and encouragement of this endeavor. I am grateful to Dr. Jonathan Wren (OMRF) for a word of advice of a seasoned editor that came in very handy during a final stage of the manuscript preparation.

I would like to acknowledge and thank my daughter Mariya Gusev (St. Petersburg Review, New York), for highly professional technical editing and for the initial cover design. Finally, my many thanks go to my publisher Mr. Stanford Chong, and to Mr. Rhaimie B. Wahap and production team of the Pan Stanford Publishing for bringing the idea behind this book to life.

Yuriy Gusev
Oklahoma City, Oklahoma, USA

CONTENTS

Preface iii

1. Measurement and Interpretation of microRNA Expression Profiles 1
   Bo Curry and Robert Ach

2. MicroRNA Expression Analysis by LNA Enhanced Microarrays 23
   Rolf Søkilde, Bogumil Kaczkowski, Kim Bundvig Barken, Peter Mouritzen, Søren Møller and Thomas Litman

3. Analysis and QC for Real-Time QPCR Arrays of Human microRNAs 47
   Dirk P. Dittmer, Pauline Chugh, Dongmei Yang, Matthias Borowiak, Ryan Ziemiecki and Andrea J. O'Hara

4. Massively Parallel microRNA Profiling in the Haematologic Malignancies 71
   Ryan D. Morin, Florian Kuchenbauer, R. Keith Humphries and Marco A. Marra

5. Who dunit? microRNAs involved in Prostate Cancer 95
   Chang Hee Kim

6. Intronic microRNA: Creation, Evolution and Regulation 117
   Sailen Barik and Titus Barik

7. miRome Architechture and Genomic Instability 133
   Konrad Huppi, Natalia Volfovsky, Brady Wahlberg, Robert M. Stephens and Natasha J. Caplen

8. Interconnection of microRNA and Transcription Factor-Mediated Regulatory Networks 149
   Yiming Zhou

9. Analysis of microRNA Profiling Data with Systems Biology Tools 169
   Yuriy Gusev, Thomas D. Schmittgen, Megan Lerner and Daniel J. Brackett

10. Mathematical and Computational Modeling of Post-Transcriptional Gene Regulation by microRNAs 197
    Raya Khanin and Desmond J. Higham

Glossary 217

Index 255

CHAPTER 1

MEASUREMENT AND INTERPRETATION OF MIRNA EXPRESSION PROFILES

Bo Curry and Robert Ach

Agilent Laboratories, Agilent Technologies
5301 Stevens Creek Blvd., Santa Clara, CA 95051

The growing recognition of the importance of miRNAs in cellular regulation has given rise to considerable interest in measuring miRNA levels in biological samples. We discuss here some of the important factors to be considered to ensure accurate, sensitive, and reproducible measurements of miRNA levels, with particular attention to the Agilent miRNA microarray system. We compare and contrast microarray with qPCR measurements, showing that there is in general excellent agreement between Agilent microarray and TaqMan qPCR miRNA measurements, though with some particular exceptions. We discuss methods of assessing RNA quality, and the importance of standardizing sample prep methods to increase reproducibility and avoid some sources of systematic variation. When choosing techniques and transforms to apply to the data, it is important to understand the sensitivity and dynamic range of the measurement. We show examples of dose-response curves for array and qPCR assays. We then discuss some potential sources of measurement bias, and propose some methods of normalizing the raw data which attempt to correct for bias from different sources. Finally, we give a brief overview of methods for analyzing miRNA expression profiling data and some caveats that must be considered.

1. Background

Many studies are revealing the complexity of the roles miRNAs play in normal and abnormal animal cell development, differentiation, and regulation (for examples of recent reviews, see[1-5]). Many of the recent findings involve measurements of miRNA expression levels. It is important for such miRNA measurements to be reproducible and quantitatively comparable from sample to sample. In a typical experimental design, cells from blood, tissue, or cell culture are isolated, their RNA is extracted and quantitated, and the amount of each targeted miRNA in the RNA sample is measured. The measured amount of each miRNA is then normalized to some measure of the original sample amount, often to a proxy for the cell count of the sample.[6] Statistical techniques are then applied to the data, often in conjunction with clinical or other expression data, to answer the research questions addressed in the experiment. Each of these steps adds statistical noise and potential bias to the results, and it is important to understand these sources of error in order to choose appropriate statistics for evaluating observed correlations and variations.

In this chapter, we discuss some of the important considerations for making accurate, reproducible miRNA measurements. We compare microarray and qPCR measurements, and discuss the importance of sample prep methods and RNA quality. We then discuss normalization methods. Appropriate normalization of raw microarray data can reduce bias in comparisons among different samples. Once measurement noise has been carefully characterized and RNA quality is well controlled, most residual variation in miRNA profiles is either due to biological noise or to the biological phenomenon that is of interest in the study.

2. Methods for Measuring miRNAs

Several methods for miRNA profiling are currently in common use. Three of the most commonly used methods are microarrays (reviewed in[7-9]), quantitative RT-PCR (qRT-PCR, or qPCR)[10-14], and high-throughput sequencing.[15] These techniques have different strengths and weaknesses, and thus are often used at different stages of the same study. Microarrays are convenient for profiling large numbers of previously characterized miRNAs. qPCR methods are well-suited to measuring smaller numbers of targets, especially in large numbers of samples. High-throughput sequencing is most suited to discovery, since it is the only method which can detect previously unknown miRNAs. We discuss below the characteristics of microarray measurements and of qPCR measurements, and a cross-platform comparison of microarrays and qPCR methods using Agilent miRNA microarrays and Applied Biosystems TaqMan qPCR.

2.1. *Microarray measurements of miRNAs*

A number of different microarray platforms have been used to profile miRNAs (for examples, see[16-24]). These platforms differ in probe design strategies and RNA labeling methods. The microarray data discussed here was acquired on the Agilent miRNA platform, following the manufacturer's recommended protocols.[25] The Agilent microarray platform features the direct end-labeling and profiling of mature miRNAs directly from total RNA without the use of size fractionation or RNA amplification.[16, 26, 27] The labeling reaction involves ligation of one labeled cytosine to the 3' end of each RNA sequence in the sample. The labeling is performed under denaturing conditions, ensuring a high labeling yield, minimal sequence bias,[16] and consistently reproducible efficiency for every miRNA sequence.[26,27]

Agilent microarray probe design features base-pairing with the additional nucleotide incorporated at the 3'-end of the miRNA during labeling.[16] Probes are designed using empirical melting point-determination, making the platform capable of single-nucleotide discrimination in the miRNA sequences.[16] Hairpin structures incorporated at the 5' end of the probes allow the binding of the mature miRNAs while discouraging the binding of longer RNAs in the total RNA sample.[16,26,27] The labeled sample is hybridized to the microarray under conditions that approach equilibrium, with a substantial and reproducible fraction of the labeled targets hybridized to the array.[16]

Agilent arrays comprise up to four different sequences probing each target miRNA, each replicated four or more times. They also include a large number of negative controls, which are designed to not hybridize to any labeled sequences, and are used for background subtraction and estimation of background noise. After hybridization, washing, and scanning, the image is processed as follows.[28] First the array features are located, and the mean signal for each feature is computed as the robust average of the counts per pixel reported for central pixels of the feature. Background fluorescence is estimated by a robust RMS fit of a surface through the negative control features, along with other features reporting weak signals not significantly higher than the average of the negative controls. The height of this fitted surface at each feature location is used to estimate the background for that feature, and the standard deviation of the residuals from the surface fit is used to estimate the noise in the background. This background is then subtracted from the mean signal of each feature, producing the background-subtracted signal (BGSubSignal). Since a large fraction of the miRNA genes are not expressed in a typical sample, the distribution of background-subtracted signals is expected to appear as a roughly Gaussian distribution centered near zero, and tailing towards the high end, where genes expressed at low levels contribute to the distribution (Figure 1).

The BGSubSignals of replicated features with the same probe sequence are then summed, after applying a statistical test to reject outliers (at $p < .05$). The sum of BGSubSignals for each probe sequence is multiplied by the number of pixels in each feature and by a scaling constant, and the product is reported as the

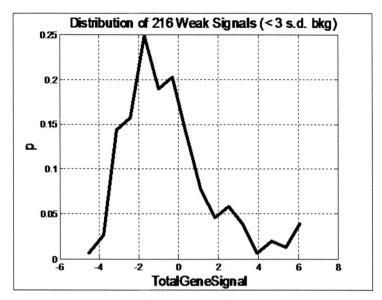

Figure 1. Distribution of weak signals after background subtraction of a typical array measurement. The width of this distribution is a measure of the background noise on the array, which determines the detection limit.

TotalProbeSignal for that probe. Next, the TotalProbeSignals for all the different probes targeting a given miRNA gene are summed to produce the TotalGeneSignal for that gene, which is reported in the GeneView summary file. The TotalGeneSignal is proportional to the total number of labeled sequences hybridized to the probes targeting each miRNA. Details of the algorithms used to accomplish these operations are available.[28]

The sum of signals of all features targeting a miRNA gene is reported, rather than their average, for two main reasons. First, it is not obvious how to appropriately average the signals from probes of varying sequences, which are expected to have significantly different sensitivities to the target miRNA. Second, for a particular array design, the sum of signals is proportional to their unweighted average; however, array designs can change on a regular basis, tracking the evolving consensus about which miRNA sequences are important.[29]

When array designs change, the TotalGeneSignal is more stable than an average. For example, 369 genes on the Agilent Human V2 array design (P/N G4470B) are targeted by the same probe sequences as those on the older Human V1 array design (P/N G4470A), but there are only 16 array features targeting each gene on the newer design (V2), rather than 20 as on the older design (V1). When 100ng of total RNA from brain or placenta were hybridized to the V1 and V2 array designs, the 75th percentile of the mean BGSubSignals reported for these 369 genes increased from 89 to 104 in brain, and from 164 to 198 in placenta. These values are consistent with the hypothesis that nearly the same amount of labeled target is hybridizing to 80% of the number of features, thereby proportionately increasing the average. By contrast, the 75th percentile of the TotalGeneSignal reported for these genes remained nearly unchanged, decreasing slightly from 116 to 109 for brain and 211 to 207 for placenta (unpublished data courtesy of Petula D'Andrade and Stephanie Fulmer-Smentek, Agilent Technologies).

The linear dynamic range and detection limit of an assay can be assessed from the response to a series of standard analytes of known concentrations. Figure 2 shows Agilent miRNA microarray results for 67 synthetic miRNA sequences, serially diluted and hybridized (data courtesy of Hui Wang, Agilent Laboratories). Nearly all sequences show a linear dynamic range of five logs. The linear dynamic range is limited at the low end by background noise on the array, and at the high end by partial saturation of the binding sites on the array. Such titration curves can be used to measure the absolute sensitivity of the assay to each miRNA target.[30] None of the miRNAs measured in this dataset exhibit high-end saturation with 1 fmol input levels, and most are detected above background at 10-100 zmol input.

As with other linear measurements, proper background subtraction is essential to maintaining response linearity at low signal levels.[31] Under-subtracting background causes low-end curvature in the signal response plot. Since there are often many miRNAs not expressed in a given sample, many miRNAs report signals less than zero after background subtraction, which can cause difficulties if the data are log-transformed. In Figure 2, miRNAs not detected above background are omitted from the plotted data. If missing data is problematic for the downstream analysis algorithms, genes expressing below background can be *surrogated*, by setting their TotalGeneSignal equal to the error in the TotalGeneSignal estimated from the error model.

In order to estimate the significance of measured expression levels and fold changes, it is necessary to associate error bounds with the reported signals. The error in the BGSubSignal of each feature is estimated using a variation on the Rosetta error model,[32,33] which includes an additive error term and a multiplicative error term.[31] The additive error is calculated from the robust standard deviation of the background (i.e. the width of the distribution in Figure 1). Since the multiplicative error term models inter-array variance, it cannot be robustly estimated for a single array.[32] Agilent recommends using 0.08 (8% error), estimated from multiple replicate experiments conducted in different

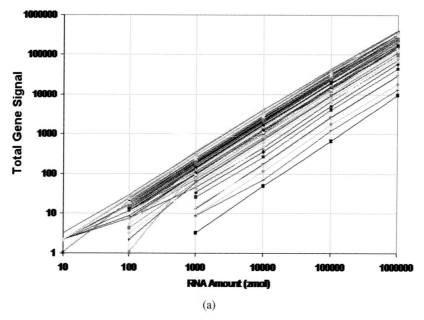

(a)

Figure 2(a). Titration of 67 synthetic miRNA sequences from 10 zmol to 1 fmol. The TotalGeneSignal was computed as described in the text. Genes reporting signals below background are omitted from the plot. Samples were prepared by serial dilutions of a labeled mixture of oligonucleotides, and hybridized to separate arrays. Data courtesy of Hui Wang, Agilent Technologies.

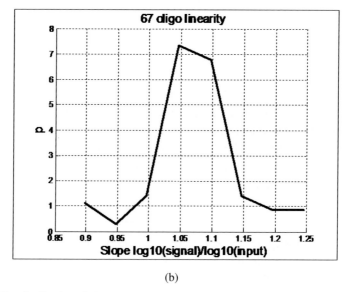

(b)

Figure 2(b). The distribution of slopes of regression lines fitted to the titration curves in A. A slope of 1.0 in a log-log plot is indicative of a linear response. Unpublished data courtesy of Hui Wang.

laboratories.[28] The feature errors are summed in quadrature to compute the TotalProbeError associated with the TotalProbeSignal, and further summed to compute the TotalGeneError, which is reported in the GeneView file.

The error model is intended to estimate the expected variation among technical replicates; it does not account for biological variation among samples, which can vary from study to study. Also, this error model does not include probe-specific or sequence-specific corrections, which could be useful in some contexts.

The quality of the array hybridization and wash can be assessed by a variety of quality metrics reported for each array.[28] For example, the additive error in the error model, which determines the lower limit of detection, can be sensitive to non-specific fluorescent background. The fraction of replicated features rejected as outliers and the coefficient of variation among replicated features not rejected are measures of the uniformity of the signals on the individual features across the array.

2.2. qPCR measurements of miRNAs

Several quantitative RT-PCR (qPCR) platforms for the measurement of miRNAs are currently commercially available. These include qPCR involving stem-loop RT primers combined with TaqMan PCR (Applied Biosystems) analysis,[10,11] qPCR to detect a ligation product using locked nucleic acid primers (Exiqon),[12] and qPCR to detect a polyadenylation product using intercalating dyes specific for double-stranded DNA (QIAGEN, Stratagene)[13,14] We have tested two of these platforms: the TaqMan and Stratagene assays. The TaqMan and Stratagene assays have both been shown to be highly specific and to have a large linear dynamic range.[10, 13]

qPCR measurements are inherently $\log_2$ measurements. In the exponential phase of the assay, each cycle represents a doubling of the signal. The cycle threshold (Ct) is defined as the cycle at which a statistically significant increase in fluorescence is detected.[34] The Ct reported by the assay, after addition of a sequence-specific constant reflecting the sensitivity of the assay to the target miRNA, is equal to the negative $\log_2$ of the initial target concentration.[35]

2.3. Comparison of qPCR and microarray measurements

Since miRNA microarray and qPCR measurements are often used in tandem, it is important to know how well the different measurements agree with each other. We recently published a detailed comparison between Agilent miRNA microarray profiles and TaqMan qPCR (Applied Biosystems) analysis,[36] which showed quantitative concordance for most miRNA targets. In this study we examined 60 different miRNAs, which were chosen to (1) represent a wide range of expression levels, (2) represent a wide range of GC content (ranging from 23-68%), and (3) include miRNAs which were shown to behave problematically in previous studies.[16,36,37] Aliquots of the same total RNA samples were used for both the array and qPCR measurements, with 100ng of total RNA input into the array labelings and 5ng input into the qPCR reverse transcription reaction. Since relative levels of miRNAs were being measured, as opposed to absolute levels, standard curves were not utilized.

Since the Ct reported by a qPCR assay measures the $-\log_2$ of the miRNA quantity (with a constant offset), both the qPCR Ct value and the $\log_2$ of the microarray signal change by a value of 1 for every 2-fold change in miRNA concentration. Therefore the most straightforward method of comparing qPCR with microarray results for a single miRNA is to plot the qPCR Ct value versus the $\log_2$ of the array signal. Figure 3 shows representative plots for nine miRNAs of Taqman qPCR Ct vs. $\log_2$(signal) from Agilent microarray analysis. If the qPCR and microarray measurements are equivalent, the plots should show a linear correlation ($R = -1$) with a slope of -1.

Figure 4 shows the slopes and the correlation values for each of 60 miRNAs tested. 56 of the 60 miRNAs show correlation values (R) between -0.8 and -1.0, and 50/60 plots have slopes between -1.2 and -0.8.

To examine results for all 60 miRNAs on one plot, one cannot simply plot qPCR Ct values versus microarray signals for all miRNAs in all tissues, because both the qPCR and microarray assays have different sensitivities to different miRNAs. To judge the consistency of fold-changes measured by microarray and qPCR platforms, we plotted the ratios of miRNA expression between all 36 possible pairs of tissues as measured by qPCR (Ct(tissue1)-Ct(tissue2)) and by microarrays ($\log_2$(signal in tissue1)-$\log_2$(signal in tissue2)). Four such plots are shown in Figure 5. The distribution of R-values for all such comparisons shows generally good correlation between the qPCR and array ratios, ranging from -0.98 to -0.82 and the distribution of slopes shows good co-linearity, varying between -1.05 and -0.79 (data not shown). The intercepts of these fold-change plots indicate the consistency between fold-changes measured by the two methods. The mean of the intercepts of the line fits for the 36 tissue pairs tested was 0.00 +/- 0.23 (1 SD).[36] This level of variability (17%) is comparable to that seen among independent measurements using the same technique.[36]

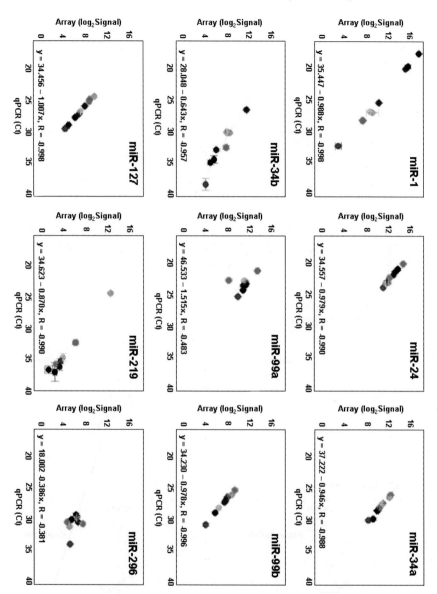

Figure 3. Representative plots for nine miRNAs of Taqman qPCR Ct vs. log$_2$(signal) from Agilent microarray analysis. Each data point represents one tissue. All plots are drawn to the same scale. The equations and R values on each plot are for the orthogonally-fitted line. Spot colors indicate the tissue: red = breast, pink = testes, dark blue = heart, light blue = placenta, dark green = liver, light green = ovary, orange = brain, brown = skeletal muscle, and grey = thymus. Tissues where qPCR results were flagged as "undetermined" by ABI software, or where log$_2$ of the total gene signal on arrays was <1, were not plotted. Error bars indicate standard deviation (SD) of Ct values for qPCR results and (SD/Mean)*log$_2$e of the signals for the array results. Figure previously published in BMC Biotechnology[36] and used with permission of the authors.

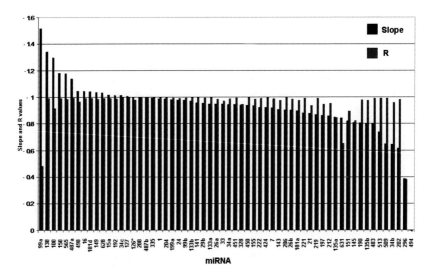

Figure 4. Slopes and the correlation values for each of 60 miRNAs tested plotted by order of the slope value. Slopes are in blue and R values are in red. Figure previously published in BMC Biotechnology[36] and used with permission of the authors.

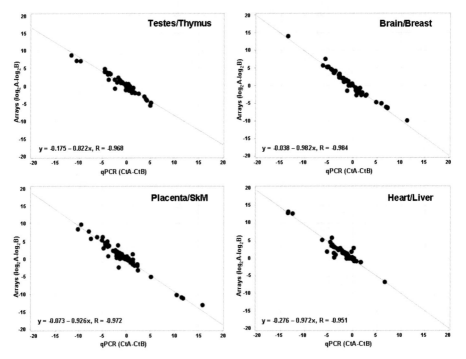

Figure 5. Ratios of miRNA expression between all 36 possible pairs of tissues as measured by qPCR (Ct(tissue1)-Ct(tissue2)). Figure previously published in BMC Biotechnology[36] and used with permission of the authors.

Though agreement between the methods is generally excellent, a few miRNAs consistently differed. miR-494 and miR-296 exhibited two of the most discrepant correlation values and slopes between the TaqMan qPCR and Agilent microarray assays. For miR-494 and miR-296, if one platform measured levels of these miRNAs accurately, while the other platform did not, then we might expect a significant divergence from linearity to be observed between the two measurements when adding increasing amounts of synthetic miR-494 or miR-296 RNA into a total RNA sample. To test this, we added 1 zmol to 10 fmol of synthetic miR-296 and miR-494 RNAs to 100ng of total RNA from liver or placenta, and measured the qPCR and microarray responses (Figure 6).

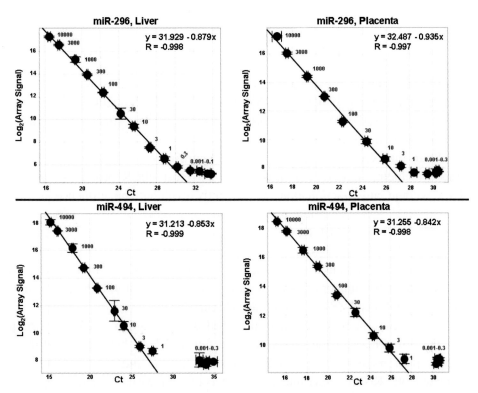

Figure 6. Scatter plots are shown for titration of synthetic miR-296 into liver (top left panel) and placenta (top right panel) total RNAs, and miR-494 into liver (lower left) and placenta (lower right) total RNAs. Ct values from qPCR are plotted on the x-axis, while $\log_2$ of the total gene signal from microarray measurements are plotted on the y-axis. Numbers in red show the number of attomoles of spike-in miRNA per 100ng total RNA. Equations and R values are for the orthogonal line fit of the linear regions of each titration. Error bars indicate standard deviation (SD) of Ct values for qPCR results and (SD/Mean)*$\log_2 e$ for the array results. Figure previously published in BMC Biotechnology[36] and used with permission of the authors.

For both miRNAs, in both tissues, the relation between qPCR measurement and array measurement was linear for four logs above a threshold spike-in concentration. The R values of the linear regions were very close to -1, with slopes between -0.842 and -0.935, indicating that both the qPCR and the microarrays were producing sample-responsive and internally consistent measurements of miR-296 and miR-494 at these concentration levels. Below the threshold spike-in levels, the qPCR Ct values and microarray signals were unchanged for miR-494, while for miR-296 the Ct values increased slightly, but the array measurements were unchanged. We conclude that the difference between the two platforms is not due to different sensitivity, since both the microarray and qPCR measurements are capable of measuring miR-296 and miR-494 accurately above a spike-in concentration threshold.

3. RNA Integrity and Quantitation

Experiments such as those described above show that these assays reproducibly measure the amount of each miRNA present in a high-quality RNA sample. However, before the RNA can be assayed it must first be extracted from various types of biological samples.

Some miRNA analysis protocols require purified small RNA fractions, while others, including the Agilent microarray and most qPCR protocols, use total RNA without any fractionation. We recently explored the influence of three popular total RNA isolation protocols on microarray profiles.[36] These were phenol/guanidinium (TRIzol, Invitrogen) followed by isopropanol precipitation, the QIAGEN miRNeasy kit, and the Applied Biosystems *mir*Vana kit. Though the measurement sensitivity for most miRNAs was unaffected by the extraction protocol, a small number of miRNAs differed systematically in the two samples we tested (HeLa and ZR-75-1 cell lines) (Figure 7).

These same changes were also observed using TaqMan qPCR, confirming that the differences in the levels of these few miRNAs between the different sample prep methods reflect true differences in the miRNA content of the extracted total RNAs, rather than just artifacts of the measurement technique. It is not known if this finding will hold true in other cell lines, or in tissue samples. Nonetheless, these results confirm the importance of holding all aspects of the measurement as constant as possible over the course of an experiment.

The integrity or degradation of total RNA is typically measured using either gel analysis or the Agilent bioanalyzer[36, 38-40] to determine the size distribution of RNA in the sample. The RNA Integrity Number (RIN) reported by the

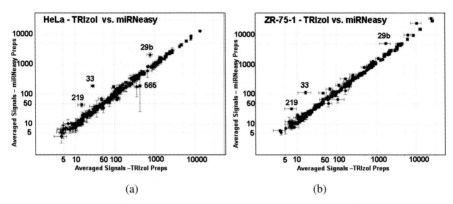

Figure 7(a)-(b). Pair-wise comparisons of averaged profiles of TRIzol and miRNeasy prep types for (A) HeLa cells and (B) ZR-75-1 cells. Total gene signals for each miRNA for all hybridizations of the same RNA prep (hybridization replicates) were averaged, and then these averaged individual prep profiles for all the preps of the same prep type were averaged together to get a mean profile for each prep type. Scatter plots show these averaged profiles from one prep type plotted against another for HeLa and ZR-75-1 cells. Error bars indicate one standard deviation. Numbers indicate the identity of all miRNAs whose signal strengths are at least two-fold higher in one prep type than another, after normalization. Figure previously published in BMC Biotechnology[36] and used with permission of the authors.

bioanalyzer provides a good measure of any large scale fragmentation of the RNA.[38, 39]

Generally the RNA sample is quantitated by UV absorption spectroscopy after purification, and RNA quantity is determined from the absorbance at 260 nm. Since both the qPCR and the Agilent microarray assays are optimized for a particular amount of total RNA input, accurate RNA quantitation is critical to obtaining reproducible results. The presence of DNA or other organic material that absorbs at or near 260nm can lead to inaccurate quantitation.[36] This can result in an over-estimation of the amount of RNA in the sample. The presence of contaminants is often estimated by looking at the 260:230 and 260:280 ratios. However, it is useful to look at the entire spectrum and not just the numerical ratio values in order to assess the presence of contaminants.[36] For specific recommendations on RNA quality, refer to platform manufacturer instructions (for example, see[25]).

4. Normalization of fold-change measurements

When comparing expression profiles of two or more samples, we often want to identify genes whose expression levels differ among the samples. The purpose of normalization is to remove the effects on signal levels of systematic differences between the measurements which are not related to the relative expression levels. Ideally, we would like to normalize to the number of cells contributing RNA to each sample; that is, the cells would be counted, the miRNA in each sample quantitatively measured, and the change in copies/cell of each miRNA directly computed. This is usually impractical, however. Instead, normalization schemes generally (a) use some internal measurement as a proxy for cell count, and (b) correct for possible differences in sensitivity between the measurements on the two samples.

It is helpful to divide the normalization task into these two independent subtasks, which are often conflated. First, normalization must select some proxy relating the number of copies of the target gene in the sample to the property of biological interest, such as the number of cells.[6] This linear normalization, in which all the signals reported in an assay are multiplied by a constant, presents the challenge of determining the appropriate constant. Second, normalization must correct for any systematic differences in the response of the assay that depend on the signal intensity, the experimental conditions, or other components of the samples. In general, this is a non-linear normalization, in which signals from different genes are scaled by different factors.

4.1. *Normalization of input RNA quantity*

miRNA expression profiles generated by qPCR or microarrays are *de facto* normalized to the amount of total RNA, since equal amounts of total RNA are used for each assay. However, accurate quantitation of total RNA in a sample is often difficult, especially for clinical samples. Thus, it is often necessary to normalize the signals from each assay by multiplication by a constant, to correct for the uncertainty in the amount of RNA input. This should be done with care, since any manipulations of the data can potentially introduce artifacts into the measurements.

The usual approach to determining an appropriate scaling constant is to select *a priori* an internal proxy for cell count, and scale the signals from each assay such that this proxy has a constant scaled signal in all assays.[6] Unfortunately, there is at present no consensus on a suitable proxy for miRNA data. In mRNA profiling it is usually assumed that either the majority of genes, or the average of all genes, or a preselected set of so-called "housekeeping" genes, can be considered to be expressed equally by all samples, and these genes are therefore used to normalize the signals of other genes.[41,42]

The housekeeping gene approach needs to be carefully applied in specific experimental data sets as there has been no universally applicable set of housekeeping genes identified for mRNA.[43] For miRNA there are too few miRNAs for majority normalization schemes such as rank consistency[28] to be statistically robust. One study reported two miRNAs which were stable and consistent normalization candidates across 18 different samples, and performed better than the usual proxies for miRNAs such as 5S RNA and U6 RNA;[41] however it is unknown whether these can be applied to all types of samples. So the choice of a proxy for normalization is still largely at the discretion of the experimenter.

A common proxy for RNA input amount is the total or average measured signal. Rank-based measures of expression levels (i.e. the signal from the Nth-highest gene) are generally more robust than total or mean signal, because they are less sensitive to very high expression of a small number of genes. The choice of the percentile to use for normalization depends on the number of genes expressed in the samples – the average expression at the chosen percentile should be well above the background noise. We have previously recommended normalization to the 75th percentile of TotalGeneSignal.[28] Normalization to a different percentile rank may also be appropriate, depending on the number of genes expressed in the samples under study. When signals are scaled, the estimated errors for those signals must be scaled by the same factor.

Some microarray assays for mRNA analysis have utilized spiked-in DNA or RNA at various points in the protocol to monitor losses or protocol failures (for example, see[44]). Such spike-ins can be useful for quality control of the assay, but cannot reliably be used for normalization, because their input amounts are fixed and not linked to any errors in the quantitation of the sample.

4.2. *Normalization of non-linear assay response*

As is clear from the foregoing discussions of array and qPCR measurements, the different assays have different sensitivities for different miRNA genes. Even what is nominally a replication of the same assay can show systematic variation due to biases in RNA extraction and cleanup, changes in laboratory protocols, or even to changes in reagents. These systematic variations are largely eliminated when differential measurements are made, and results are reported as fold changes between samples measured with the same assay and protocol.

Though the response of a qPCR or array assay is ideally linearly dependent on the input quantity (e.g. Fig. 2), this is sometimes not the case. Several methods have been proposed to correct for intensity-dependent nonlinearities, which are observed on some platforms.[45,46] Also, the sensitivity of the assay to different genes can vary with changes in protocols, reagents, equipment, or personnel, as mentioned in Section 3. Such systematic condition-dependent nonlinearities can potentially be minimized by normalization to a control or reference sample.[47] Finally, assay sensitivity can vary from sample to sample due to interactions of the genes, probes, or PCR primers with other components of the complex sample. Such sample-dependent non-linearities cannot be corrected for using internal evidence from a single experiment, but can often be identified using a series of standard additions[48] (e.g. Fig. 6). The effects of sample-dependent non-linearities on measured fold changes can be minimized by normalization to a near-reference sample, as close as possible to the test samples in overall RNA expression.[47]

All such non-linear normalization schemes add some risk of artificially distorting expression profiles, and their effectiveness depends on the goals and design of the experiment. Details about the appropriate use of these methods in specific situations are a subject of continuing research and debate in the community, and are beyond the scope of this chapter.

4.3. *Avoiding normalization entirely*

Linear normalization and other methods for intensity-dependent normalization in common use adjust the magnitude of reported signals without changing their ranks. One way to obviate some normalization issues is to eschew normalization entirely, and use rank-based statistics to decide which genes are differentially expressed.[49] Rather than identifying genes whose expression levels differ significantly, such schemes identify genes whose ranks, relative to other expressed genes, change significantly. Obviously, information about the magnitude of fold changes is lost with this approach. The detection of significantly changed expression may, however, be more robust, and some normalization artifacts may be avoided.

5. Data Analysis

To gain a better understanding of miRNA activities and to identify their diagnostic potential, scientists use a variety of data analysis tools to interpret and analyze miRNA profiling data. High-throughput measurement is often used in sample cohorts to identify profile differences when comparing different biological characteristics of the samples. Pioneering work of this type was reported for mRNA expression profiling, such as in the analysis of differential expression between two types of leukemia.[50]

The basic level analysis of miRNA expression data is in many ways very similar to the analysis of mRNA expression data. Standard tools such as computing differential expression and clustering are used in a very similar manner. Data analysis methods for mRNA microarray expression data have been described in many studies (reviewed in[51]). The application of such methods to miRNA data was demonstrated in early papers such as Lu et al.[52], where miRNA differential expression in cancer versus non-cancer samples was reported. Other studies have applied and extended various data analysis tools for miRNA profiles.

For example, in one study miRNA profiling was carried out for normal tissues and different types of cancer tissues.[53] After identifying the miRNAs differentially expressed for each cancer type the authors focused on identifying miRNAs differentially expressed across many cancer types. In another study miRNA profiling data was collected from different tumor samples where the primary origin was known.[54] Using a machine learning approach a classifier for cancer tissue of origin was constructed that makes use of only miRNA expression signature information.

There are a couple of important caveats that can be useful to consider when analyzing miRNA profiling data. First, there are generally fewer differentially expressed miRNAs than there are protein coding genes profiled in a typical mRNA assay. Therefore standard distribution assumptions, which depend on large numbers of data points, may not hold for miRNA data. Also, the large dynamic range of miRNA expression levels, and substantial variations in the total miRNA levels observed between some different samples, can be significant challenges to analysis.[11, 52]

miRNAs are considered master regulators in the cell, as a single miRNA can regulate numerous genes.[55] As such, the set of genes that is targeted by any given miRNA is of particular interest when studying the effects and roles of miRNAs. Several groups have developed computational tools to predict the targets of miRNAs, taking into account sequence motifs, thermodynamic principles, and conservation during evolution (for examples, see[56-59]). A major factor that determines whether a gene is targeted by a given miRNA is sequence complementarity of the miRNA seed region to sequences in the gene's 3' untranslated region.[57]

For a given miRNA of interest, target prediction tools provide numerical assessment of the interaction between the miRNA and potential target RNA molecules. Many of the prediction tools offer databases that are accessible and frequently used by the scientific community (e.g. TargetScan[58], PITA[60] and miRBase[29]). The predicted targets have been experimentally validated in some cases by perturbing the miRNA levels and observing changes in mRNA and protein levels.[61,62] Joint analysis of miRNA expression profiles with mRNA or protein expression profiles can supplement *in silico* target prediction to suggest with more confidence candidate targets for further validation.[63] Target information can be incorporated into the analysis of miRNA expression data to gain better understanding of the biological pathways that may be associated with differences in expression.

In summary, there are several types of data analysis and visualization tools usefully employed for analysis and interpretation of miRNA profile data. We expect data analysis methods to continue to develop and evolve, enabling further insights about the detailed mechanisms by which miRNAs function in normal and abnormal cell development, differentiation, and regulation.

Acknowledgements

We thank Hui Wang, Petula D'Andrade, and Stephanie Fulmer-Smentek for generous use of their unpublished data. We thank Hui Wang, Stephanie Fulmer-Smentek, Laurakay Bruhn, Steve Laderman, Jeff Sampson, Becky Mullinax, Israel Steinfeld, and Zohar Yahkini for their comments and suggestions on the manuscript.

Part of this manuscript, including Figures 3-7, was previously published in BMC Biotechnology by Robert Ach, Hui Wang, and Bo Curry (Ach, R; Wang, H; Curry, B: Measuring microRNAs: Comparisons of microarray and quantitative PCR measurements, and of different total RNA prep methods. BMC Biotechnology 2008, 8: 69). This content was used with permission of the authors.

References

1. Schickel, R; Boyerinas, B; Park, SM; Peter, ME: MicroRNAs: key players in the immune system, differentiation, tumorogenesis, and cell death. *Oncogene* 2008, 9: 5959-5974.
2. Esquela-Kerscher, A; Slack, FJ: Oncomirs – microRNAs with a role in cancer. *Nat. Rev. Cancer* 2006, 6: 259-269.
3. Flynt, AS; Lai, EC: Biological principles of microRNA-mediated regulation: shared themes amid diversity. *Nat. Rev. Genet.* 2008, 9: 831-842.
4. Erson, AE; Petty, EM: MicroRNAs in development and disease. *Clin. Genet.* 2008, 74: 296-306.
5. Zhang, C: MicroRNomics: a newly emerging approach for disease biology. *Physiol. Genomics* 2008, 33: 139-147.
6. Kanno, J; Aisaki, K-I; Igarashi, K; Nakatsu, N; Ono, A; Kodama, Y; Nagao, T: "Per cell" normalization method for mRNA measurement by quantitative PCR and microarrays. *BMC Genomics* 2006, 7: 64.
7. Yin, JQ; Zhao, RC; Morris, KV: Profiling microRNA expression with microarrays. *Trends Biotechnol.* 2007, 26: 70-76.

8. Davison, TS; Johnson, CD; Andruss, BF: Analyzing microRNA expression using microarrays. *Meth. Enzymol.* 2006, 411: 14-34.
9. Kong, W; Zhao, J-J, and Cheng, JQ: Strategies for profiling microRNA expression. *J. Cell. Physiol.* 2008, 218: 22-25.
10. Chen, C; Ridzon, DA; Broomer, AJ; Zhou, Z; Lee, DH; Nguyen, JT; Barbisin, M; Xu, NL; Mahuvakar, VR; Andersen, MR; Lao, KQ; Livak, KJ; Guegler, KJ: Real-time quantification of microRNAs by stem-loop RT-PCR. *Nuc. Acids Res.* 2005, 33: e179.
11. Liang, Y; Ridzon, D; Wong, L; and Chen, C: Characterization of microRNA expression profiles in normal human tissues. *BMC Genomics* 2007, 8: 166.
12. Raymond, CJ; Roberts, BS; Garrett-Engele, P; Lim, LP; and Johnson, JM: Simple, quantitative primer-extension PCR assay for direct monitoring of microRNAs and short-interfering RNAs. *RNA* 2005, 11: 1737-1744.
13. Stratagene [http://www.stratagene.com/products/displayProduct.aspx?pid=820].
14. QIAGEN [http://www1.qiagen.com/products/miRNA/miScriptSystem.aspx].
15. Landgraf, P; Rusu, M; Sheridan, R; Sewer, A; Iovino, N; Aravin, A; Pfeffer, S; Rice, A; Kamphorst, AO; Landthaler, M; Lin, C; Socci, ND; Hermida, L; Fulci, V; Chiaretti, S; Foa, R; Schliwka, J; Fuchs, U; Novosel, A; Muller, R-U; Schermer, B; Bissels, U; Inman, J; Phan, Q; Chien, M; Weir, DB; Choksi, R; De Vita, G; Frezzeti, D; Trompeter, H-I; Hornung, V; Teng, G; Hartmann, G; Palkovits, M; Di Lauro, R; Wernet, P; Macino, G; Rogler, CE; Nagle, JW; Ju, J; Papavasiliou, FN; Benzing, T; Lichter, P; Tam, W; Brownstein, MJ; Bosio, A; Borkhardt, A; Russo, JJ; Sander, C; Zavolan, M; Tuschl, T: A mammalian microRNA expression atlas based on small library sequencing. *Cell* 2007, 129: 1401-1414.
16. Wang, H; Ach, RA; Curry, B: Direct and sensitive miRNA profiling from low-input total RNA. *RNA* 2007, 13: 151-159.
17. Castoldi, M; Schmidt, S; Benes, V; Noerholm, M; Kulozik, AE; Hentze, MW; Muckenthaler, MU: A sensitive array for microRNA expression profiling (miChip) based on locked nucleic acids (LNA). *RNA* 2006, 12: 913-920.
18. Castoldi, M; Schmidt, S; Benes, V; Hentze, MW; Muckenthaler, MU: miChip: an array-based method for microRNA expression profiling using locked nucleic capture probes. *Nat. Protoc.* 2008, 3: 321-329.
19. Baskerville, S; Bartel, DP: Microarray profiling of microRNAs reveals frequent coexpression with neighboring miRNAs and host genes. *RNA* 2005, 11: 241-247.
20. Liu, C-G; Calin, GA; Meloon, B; Gamliel, N; Sevignani, C; Ferracin, M; Dumitru, CD; Shimizu, M; Zupo, S; Dono, M; Alder, H; Bullrich, F; Negrini, M; Croce, C: An oligonucleotide microchip for genome-wide microRNA profiling in human and mouse tissues. *Proc. Natl. Acad. Sci. USA* 2004, 101: 9740-9744.
21. Beuvink, I; Kolb, FA; Budach, W; Garnier, A; Lange, J; Natt, F; Dengler, U; Hall, J; Filipowicz, W; Weiler, J: A novel microarray approach reveals new tissue-specific signatures of known and predicted mammalian microRNAs. *Nuc. Acids Res.* 2007, 35: e52.
22. Shingara, J; Keiger, K; Shelton, J; Laosinchai-Wolf, W; Powers, P; Conrad, R; Brown, D; Labourier, E: An optimized isolation and labeling platform for accurate microRNA expression profiling. *RNA* 2005, 11: 1461-1470.
23. Barad, O; Meiri, E; Avniel, A; Aharonov, R; Barzilai, A; Bentwich, I; Einav, U; Gilad, S; Hurban, P; Karov, Y; Lobenhofer, EK; Sharon, E; Shiboleth, YM; Shtutman, M; Bentwich, Z; Einat, P: MicroRNA expression detected by oligonucleotide microarrays: system establishment and expression profiling in human tissues. *Genome Res.* 2004, 14: 2486-2494.
24. Goff, LA; Yang, M; Bowers, J; Getts, RC; Padgett, RW; Hart, RP: Rational probe optimization and enhanced detection strategy for microRNAs using microarrays. *RNA Biol.* 2005, 2: 93-100.

25. Agilent Technologies [http://www.chem.agilent.com/Library/usermanuals/Public/G4170-90011_miRNA_Complete_2.0.pdf].
26. Agilent Technologies [https://agilenteseminar.webex.com/agilenteseminar/onstage/g.php?AT=VR&RecordingID=277729994].
27. Agilent Technologies [http://www.opengenomics.com/webcastinfo.aspx?wid=28].
28. Agilent Technologies [http://www.chem.agilent.com/Library/usermanuals/Public/G4460-90010_FE_Reference.pdf].
29. Griffiths-Jones, S; Grocock, RJ; van Dongen, S; Bateman, A; Enright, AJ: miRBase: microRNA sequences, targets and gene nomenclature. *Nuc. Acids Res.* 2006, 34: D140-144.
30. Rutledge, RG: Sigmoidal curve-fitting redefines quantitative real-time PCR with the prospective of developing automated high-throughput applications. *Nuc. Acids Res.* 2004, 32: e178.
31. Huber, W; von Heydebreck, A; Vingron, M: Error models for microarray intensities. In *Encyclopedia of Genomics, Proteomics and Bioinformatics*. Edited by Dunn, MJ; Jorde, LB; Little, PFR; Subramaniam, S. New Jersey: John Wiley & sons; 2005.
32. Weng, L; Dai, H; Zhan, Y; He, Y; Stepaniants, SB; Bassett, DE: Rosetta error model for gene expression analysis. *Bioinformatics* 2006, 22: 1111-1121.
33. Rocke, DM; Durbin, B: A model for measurement error for gene expression arrays. *J. Computational Biol.* 2001, 8:557–569.
34. Applied Biosystems [http://www3.appliedbiosystems.com/cms/groups/mcb_marketing/documents/ generaldocuments/cms_053906.pdf].
35. Kubista, M; Sindelka, R: The prime technique: Real-time PCR data analysis. *GIT Lab. J. Eur.* 2007, 9-10: 33-35.
36. Ach, R; Wang, H; Curry, B: Measuring microRNAs: Comparisons of microarray and quantitative PCR measurements, and of different total RNA prep methods. *BMC Biotechnol.* 2008, 8: 69.
37. Enos, JM; Duzeski, JL; Roesch, PL; Hagstrom, JE; Watt, M-AV: MicroRNA labeling methods directly influence the accuracy of expression profiling detection. *BioTechniques* 2007, 42:378-381.
38. Schroeder, A; Mueller, O; Stocker, S; Salowsky, R; Leiber, M; Gassmann, M; Lightfoot, S; Menzel, W; Granzow, M; Ragg, T: The RIN: an RNA integrity number for assigning integrity values to RNA measurements. *BMC Mol. Biol.* 2006, 7: 3.
39. Agilent Technologies [http://www.chem.agilent.com/en-US/Products/Instruments/lab-on-a-chip/Pages/gp14975.aspx].
40. Agilent Technologies [http://www.chem.agilent.com/en-us/products/instruments/lab-on-a-chip/2100bioanalyzer/pages/default.aspx].
41. Peltier, HJ; Latham, GJ: Normalization of microRNA expression levels in quantitative RT-PCR assays: Identification of suitable reference RNA targets in normal and cancerous human solid tissues. *RNA* 14: 844-852.
42. Vandesompele, J; De Preter, K; Pattyn, F; Poppe, B; Van Roy, N; De Paepe, A; Speleman, F: Accurate normalization of real-time quantitative RT-PCR data by geometric averaging of multiple internal control genes. *Genome Biol.* 2002, 3: research0034.1-0034.11.
43. Takagi, S; Ohashi, K; Utoh, R; Tatsumi, K; Shima, M: Okano, T: Suitable reference genes for the analysis of direct hyperplasia in mice. *Biochem. Biophys. Res. Commun.* 2008, 377: 1259–1264.
44. Agilent Technologies [http://www.chem.agilent.com/Library/usermanuals/Public/Spike-in%20Kit.pdf].

45. Bolstad, BM; Irizarry, RA; Astrand, M; Speed, TP: A comparison of normalization methods for high density oligonucleotide array data based on variance and bias. *Bioinformatics*, 2003, 19: 185–193.
46. Workman, C: Jensen, LR; Jarmer, H; Berka, R; Gautier, L; Nielsen, HB; Saxild, H-H; Nielsen, C.;Brunak, S; Knudsen, S: A new non-linear normalization method for reducing variability in DNA microarray experiments. *Genome Biol.* 2002, 3: research0048.1–0048.16.
47. He, YD; Dai, H; Schadt, EE; Cavet, G; Edwards, SW; Stepaniants, SB; Duenwald, S; Kleinhanz, R; Jones, AR; Shoemaker, DD; Stoughton, RB: Microarray standard data set and figures of merit for comparing data processing methods and experiment designs. *Bioinformatics* 2003, 19: 956–965.
48. Harris, C: *Quantitative Chemical Analysis*. 6th Edition. New York: W.H. Freeman; 2003.
49. Ben-Dor, A; Friedman, N; Yahkini, Z: Scoring Genes for Relevance. *Technical Report* 2000-38; Jerusalem: Institute of Computer Science, Hebrew University; 2000 [http://leibniz.cs.huji.ac.il/tr/340.ps].
50. Golub, TR; Slonim, DK; Tamayo, P; Huard, C; Gaasenbeek, M; Mesirov, JP; Coller, H; Loh, ML; Downing, JR; Caliqiuri, MA; Bloomfield, CD; Lander, ES: Molecular classification of cancer: class discovery and class prediction by gene expression monitoring. *Science* 1999, 286: 531-537.
51. Causton, HC; Quackenbush, J; Brazma, A: *Microarray Gene Expression Data Analysis: A Beginner's Guide*. New Jersey: Blackwell Publishing; 2003.
52. Lu, J ; Getz, G ; Miska, EA ; Alvarez-Saavedra, E ; Lamb, J ; Peck, D ; Sweet-Cordero, A; Ebert, BL; Mak, RH; Ferrando, AA; Downing, JR; Jacks, T; Horvitz, HR; Golub, TR: MicroRNA expression profiles classify human cancers. *Nature* 2005, 435: 834-8.
53. Volinia, S; Calin, GA; Liu, CG; Ambs, S; Cimmino, A; Petrocca, F; Visone, R; Iorio, M; Roldo, C; Ferracin, M; Prueitt, RL; Yanaihara, N; Lanza, G; Scarpa, A; Vecchione, A; Negrini, M; Harris, CC; Croce, CM: A microRNA expression signature of human solid tumors defines cancer gene targets. *Proc. Natl. Acad. Sci. USA* 2006, 103: 2257-2261.
54. Rosenfeld, N; Aharonov, R; Meiri, E; Rosenwald, S; Spector, Y; Zepeniuk, M; Benjamin, H; Shabes, N; Tabak, S; Levy, A; Lebanony, D; Goren, Y; Silberschein, E; Targan, N; Ben-Ari, A; Gilad, S; Sion-Vardy, N; Tobar, A; Feinmesser, M; Kharenko, O; Nativ, O; Nass, D; Perelman, M; Yosepovich, A; Shalmon, B; Polak-Charcon, S; Fridman, E; Avniel, A; Bentwich, Z; Cohen, D; Chajut, A; Barshack, I: MicroRNAs accurately identify cancer tissue origin. *Nature Biotech.* 2008, 26: 462-469.
55. Lim, LP; Lau, NC; Garrett-Engele, P; Grimson, A; Schelter, JM; Castle, J; Bartel, DP; Linsley, PS; Johnson, JM: Microarray analysis shows that some microRNAs downregulate large numbers of target mRNAs. *Nature* 2005, 433: 769-73.
56. Krek, A; Grun, D; Poy, MN; Wold, R; Rosenberg, L; Epstein, EJ; MacMenamin, P; da Piedade, I; Gunsalus, KC; Stoffel, M; Rajewsky, N: Combinatorial microRNA target predictions. *Nat. Genet.* 2005, 37: 495-500.
57. Lewis, BP; Shih, I-H; Jones-Rhoades, MW; Bartel, DP; Burge, CB: Prediction of mammalian microRNA targets. *Cell* 2003, 115: 787–798.
58. Lewis, BP; Burge, CB: Bartel, DP: Conserved seed pairing, often flanked by adenosines, indicates that thousands of human genes are microRNA targets. *Cell* 2005, 120: 15-20.
59. Miranda, KC; Huynh, T; Tay, Y; Ang, Y-S; Tam, W-L; Thomson, AM; Lim, B; Rigoutsos, I: A pattern-based method for the identification of microRNA binding sites and their corresponding heteroduplexes. *Cell* 2006, 126: 1203-1217.
60. Kertesz, M; Iovino, N; Unnerstall, U: Gaul, U; Segal, E: The role of site accessibility in microRNA target recognition. *Nat. Genet.* 2007, 39: 1278-1284.

61. Selbach, M; Schwanhausser, B; Thierfelder, N; Fang, Z; Khanin, R: Rajewsky, N: Widespread changes in protein synthesis induced by microRNAs. *Nature* 2008, 455: 58 – 63.
62. Baek, D; Villen, J; Shin, C; Camargo, FD; Gygi, SP; Bartel, DP: The impact of microRNAs on protein output. *Nature* 2008, 455: 64-71.
63. Creighton, CJ; Nagaraja, AK; Hanash, SM; Matzuk, MM; Gunaratne, PH: A bioinformatics tool for linking gene expression profiling results with public databases of microRNA target predictions. *RNA* 2008, 14: 2290-96.

CHAPTER 2

MICRORNA EXPRESSION ANALYSIS BY LNA ENHANCED MICROARRAYS

Rolf Søkilde [1,2], Bogumil Kaczkowski [2], Kim Bundvig Barken [2], Peter Mouritzen [2], Søren Møller [2], Thomas Litman [2]

[1] *University of Copenhagen, Faculty of Health Sciences*
[2] *Exiqon A/S*

Array expression profiling of miRNAs has become a widely used technique and thereby gained a broader user spectrum from novice to highly experienced array specialists. In this chapter we discuss the challenges in designing microarrays for miRNA detection and show the advantages of LNA™-enhanced capture probes. We cover important aspects of array normalization and show its practical application in a case study with miRNA data from the NCI-60 cell line panel. We also demonstrate that commonly used methods for normalization of mRNA expression arrays can be applied for miRNA array data. These methods are implemented in both open source and commercial software.

1. Introduction

microRNAs (miRNAs) constitute an important class of non-coding, short, endogenous RNAs that act as post-transcriptional regulators of mRNAs by base pairing to their 3' untranslated regions. The mature miRNAs (19-25 nucleotides long) are processed from longer, hairpin transcripts by the RNAse III ribonucleases Drosha1 and Dicer.[1] The miRBase database at the Welcome Trust Sanger Institute annotates microRNAs based on their experimental detection, including cloning studies that define the sequence boundaries of miRNAs.[2]

Most miRNAs can be grouped in families based on their sequence similarity. At least some of these families have evolved by duplication events, such as the miR-17-92 cluster.[3] As a result, the biological effect on a target mRNA may be redundant and additive (when the miRNAs in a family are very similar), but can also result in a variety of specificities towards many different mRNAs (when the family members diverge and display differential expression in time and space).

Thus, miRNAs are believed to act as key regulatory molecules in many important biological processes, including developmental timing and differentiation of tissue,[4] as well as in many diseases such as cancer and neurological diseases.[5-7]

The first microarrays designed to detect miRNAs were mentioned as early as 2003 and contained simple DNA capture probes.[8] In 2005, microarrays that used LNA™ for melting temperature (Tm) normalization of the capture probes appeared.[9] LNA™ or Locked Nucleic Acids are modified RNA nucleotides in which the ribose moiety has been modified with an extra methylene bridge connecting the 2'-O atom with the 4'-C atom of the ribose (see Figure 1).

This bridge locks the ribose ring in its 3'-endo structural conformation, which enhances both base stacking and alignment of the backbone, thus, significantly increasing the thermal stability of the oligonucleotide duplex. LNA™-modified nucleotides follow standard Watson-Crick base pairing rules. The higher thermal stability (Tm increases by 2-8°C per LNA™ base) is ideal for the design of short probes in highly sensitive assays such as microarrays, northern blots, in situ hybridization, and qPCR. In addition, LNA™ probes enable increased mismatch discrimination, which is essential for highly specific detection of miRNAs.[10]

In this chapter we outline the performance of LNA™-enhanced arrays, and demonstrate robust single channel and dual color applications for profiling miRNA expression with the NCI-60 cell line panel.

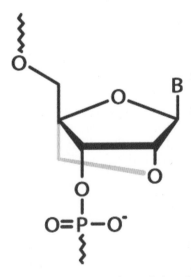

Figure 1. The structure of an LNA monomer, note the methylene bridge which locks the ribose configuration.

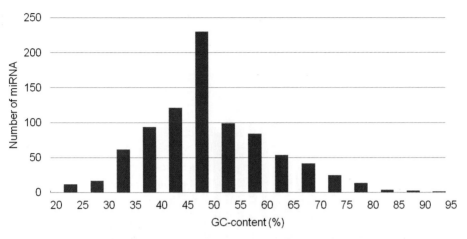

Figure 2. GC-content of mature miRNA sequences from miRBase version 12.0. X-axis shows different ranges of QC.

2. Design of Microarrays for miRNA Expression Analysis

Probe design for miRNAs is particularly challenging for at least two reasons:

- The GC content of miRNAs can vary from 16-91% (see Figure 2).
- Many miRNAs display a high degree of sequence similarity. For example, sequences with only one nucleotide difference are common, such as the miRNA-10a/b pair (see Figure 4).

The ability to predict the melting temperature, Tm, is important for successful design of probes. Tm is defined as the temperature at which half of the duplexes are denatured into single stranded molecules. The melting temperature can be predicted by different Tm models or measured experimentally.[11] The Tm of a probe indicates how willingly it will bind to its target under a given set of experimental conditions of an array experiment. A too low Tm will give low affinity for the target resulting in low sensitivity, while a Tm that is too high may increase the likelihood of unspecific binding.

The GC content of human miRNAs annotated in miRbase varies from 16-91%.[2] This means that the Tm of full length complementary DNA probes will vary from 50-80°C (see Figure 3). Given the very short length of miRNA sequences and their highly variable GC content, it is impossible to design an array of DNA capture probes with an equal Tm targeting all known miRNAs. Design of Tm equalized capture probes is essential for construction of a highly

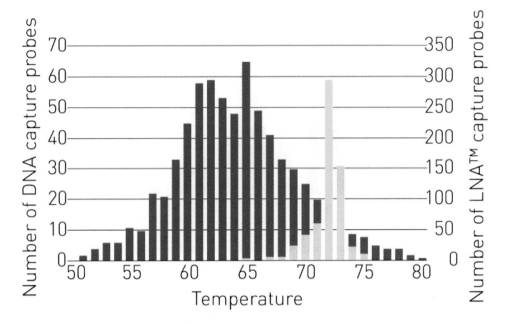

Figure 3. Comparison of the Tm distribution of capture probes when using either Tm normalized LNA capture probes in yellow or full length DNA capture probes in dark grey.

discriminative and sensitive array. By placing a few LNA™ nucleotides in specific positions in the probe sequence, it is possible to achieve a uniform Tm for all probes independent of their GC content (see Figure 3). Inclusion of LNA™ also has a dramatic effect on probe specificity shown by an increase in the discriminative power of probes designed to distinguish different miRNAs within closely related families (see Figure 4).

3. Experimental Verification of Probe Performance

The mismatch discriminating power of the array was evaluated using pools of synthetic miRNAs. The closely sequence related miRNAs were divided into 16 different pools maximizing the diversity of each pool regarding sequence similarity.

As an example of how well the probes detect the correct target the miR-10 and miR-99 families are shown in Figure 4. In the experiment the pools of

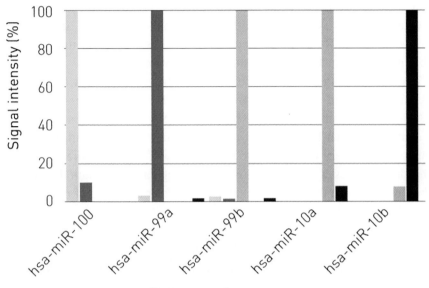

miR	Sequence
hsa-miR-100	5'-aaccc-gtagatccgaacttgtg-3'
hsa-miR-99a	5'-aaccc-gtagatccgatcttgtg-3'
hsa-miR-99b	5'-caccc-gtagaaccgaccttgcg-3'
hsa-miR-10a	5'-taccctgtagatccgaatttgtg-3'
hsa-miR-10b	5'-taccctgtagaaccgaatttgtg-3'

Figure 4. Specificity of the capture probes targeting the miR-10 and miR-99 families, which include miR-10a, miR-10b, miR-99a, miR-99b and miR-100. On the top figure the x-axis shows the miRNAs that were used in the spike-in experiment. On the y-axis the intensity of signal from the perfect matching probe is set to 100%, while the rest is calculated as % signal intensities of this. On the bottom figure the mismatches in the aligned sequences are highlighted in yellow.

synthetic miRNAs are labeled individually and hybridized to arrays. As a measure of cross hybridization the correct signal was set to 100 percent and any signal on the closely related miRNA capture probes was calculated as a percentage of this value. E.g. spiking in miR-99a results in 10% cross hybridization to miR-100, while the rest of the closely related miRNA capture probes give less than 1% cross hybridization to this miRNA.

When comparing sensitivity data obtained from arrays with capture probes containing LNA and arrays containing full length capture probes based only on DNA, it is evident that the sensitivity of the DNA probes varies depending on GC content, and that DNA capture probes targeting miRNAs with very low GC amount perform very poorly due to the very low Tm (see Figure 5). The LNA probe design gives a better sensitivity over regular full length DNA capture probes, especially in the low GC region, with greater than 90% of miRNA targets detected at 50 amol concentration for miRNA with GC content lower than 70%.

It has been demonstrated that several length variants of a particular miRNA can co-exist in cells or tissue, so-called isomiRs.[12-14] The difference in sequence between the variants is primarily found at the 3'-end of the miRNA molecule. The functional significance of this 3'-end heterogeneity is still not fully understood, but it is likely that the regulatory effect of microRNAs is not restricted to one particular variant. When analyzing microRNA expression, it is, therefore, important that all variants are detected.

The miRNA sequence annotated in miRBase will often be that of the predominant variant. However, it may not be the predominant variant in all cells and tissues and the polymorphic profile can vary from tissue to tissue.[14] The biological relevance of 3'-end heterogeneity has been discussed in several publications, and it has been suggested that it might play a role in subcellular localization or functional efficacy[13] and that untemplated heterogeneity[15]—where the miRNA no longer matches the genome in the 3' end, might be a signal for microRNA degradation.[15] Wu et al. predict that enzymes with end-modifying activity that act downstream of the Drosha/Dicer processing might be the cause of the end heterogeneity, while others suggest that it might be due to inaccurate Dicer processing or degradation of the 3'-end of the microRNA.[16,17] Morin et al. have recently also observed 3' variants where the most abundant miRNA sequence is not the miRBase reference sequence, and they notice that in later miRBase updates the more abundant forms are becoming the reference sequence.[14]

Techniques for detection of microRNAs often have different sensitivity towards such length variants because the assays, such as Q-RT-PCR, were

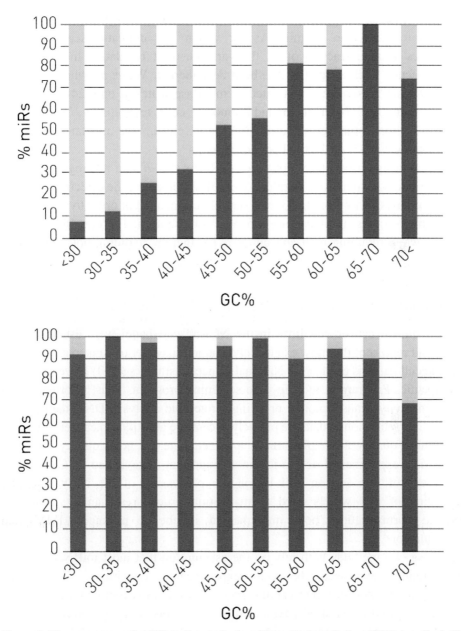

Figure 5. The percentage of miRNAs detected when 50 amol of synthetic miRNA is spiked. The experiment was conducted with 557 different synthetic miRNAs and the detection limit was set to be 5 times the standard deviation above the background intensity. The x-axis shows the GC content of the miRNAs and the y-axis is the percentage of capture probes detecting 50 amol targets within each of the GC ranges. The yellow bars show the LNA capture probe performance, while the grey bars show the DNA capture probe. It is clear that the DNA based capture probes have difficulties detecting the low GC miRNAs.

designed to specifically detect the sequence annotated in miRBase. This can be a problem for detection of miRNAs for which the annotated sequence only constitutes a small fraction of all the variants, or if regulation in the variant profile occurs. Such a situation has been described by Wu et al. who found that the mmu-miR-21 sequence annotated in miRBase matched only 24-57% of the variants found in Naive, Effector and Memory T cells, respectively.[13] In addition, Ruby et al. found that for some miRNAs in C. elegans the annotated sequence only constitutes 16% of the total pool of variants.[17]

When designing capture probes for miRNAs this 3'-end heterogeneity must be taken into consideration; therefore, careful placement of the detection probes a couple of bases away from the 3' end should be practiced in order to also capture the shorter variants. LNA capture probes can be shortened (preferably from the 3' end) while retaining high and uniform Tm thereby avoiding potential biases introduced by reduced hybridization of 3' variants.

4. MicroRNA Expression Profiling of the NCI-60 Cell Lines

The 59 human cancer cell lines in the NCI-60 cell panel were established in 1990, and represent the most extensively characterized set of cells available. The NCI-60 cell panel is used at the National Cancer Institute's (NCI) Developmental Therapeutics Program in their in-vitro screening program. Thus, information on drug sensitivity, transcriptomic profiling, proteomic analysis, DNA methylation status, and metabolomic data are publicly accessible at http://dtp.cancer.gov.[18-20]

A high quality array platform is a prerequisite for good quality data, but appropriate experimental design is just as important as reviewed by Churchill.[21] The focus of this section will be on the processing of the array data and evaluation of two widely used normalization procedures, namely loess and quantile normalization, both implemented in the R package LIMMA in Bioconductor.[22]

These two methods both rely on the assumption that most miRNAs are not differentially expressed between experimental conditions. There are approximately 3400 different capture probes on the array and of these, 984 are more than 2-fold differentially regulated in at least one cell line compared to the common reference. For the individual cell lines the number of miRNAs that are more than 2-fold differentially expressed (compared to the common reference) is between 52 and 365 miRNAs with an average of 120 miRNAs. For the most variable cell lines there is 37% of the miRNAs that are differentially expressed between the common reference and that particular cell line. The average percentage of differentially expressed miRNAs is only 12% between a cell line and the common reference channel. Thus, the assumption that most miRNAs are not differentially expressed holds for the use of loess and quantile normalization.

Other methods which could be considered in other experimental setups include variance stabilization normalization (VSN), spike-in VSN, global median and print-tip loess.[23, 24]

In this section we compare miRNA expression from the NCI-60 cell line panel using two alternative data processing approaches to a common reference design experiment:

- Two color ratio data;
- Single channel intensity data.

4.1. Cell Line Cultivation and RNA Extraction

The cell lines were grown at 37 °C to near confluence in RPMI-1640 media, supplemented with 1% L-Glutamine and 5% FBS, before total RNA was isolated by guanidinum isothiocyanate / phenol: chloroform extraction (Trizol). This RNA extraction method ensures preservation of the small RNA fraction.

4.2. Experimental setup

For analysis of the NCI-60 cell lines' miRNA profiles, a common reference design was applied as this enabled both one- and two-channel data analysis (Figure 6). The sample channel was labeled with a Hy3™ dye (green fluorescent), while the reference channel was labeled with a Hy5™ dye (red fluorescent). The common reference is composed of an equimolar mix of total RNA from all 59 cell lines (59 of the original 60 NCI-60 cell lines are presently available). A direct enzymatic labelling of the miRNA molecules was carried out according to the manufacturer's instructions (www.exiqon.com).

The common reference can be used as a common factor to which signals from all samples can be normalized, enabling direct comparison. We recommend using a common reference design when the study contains more than two different types of sample to be compared. In addition, it also has advantages to typical paired sample studies in that it allows technical and biological variation to be separated. The common reference design provides information concerning variation between replicates and, importantly, can be used to identify outlying samples.

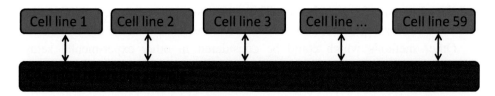

Figure 6. The common reference sample is a mixture of RNA from all the cell lines, against which the individual cell lines are compared. The reference sample is labeled with Hy5 (red fluorescent dye), while RNA from the individual cell lines is labeled with Hy3 (green fluorescent dye).

The ideal reference sample is one that is as similar as possible to the actual samples and is likely to contain all miRNAs found in any of the samples in the study. Both of these conditions are fulfilled by making a reference sample from a pool of all the samples in the study. This requires that there is enough RNA available from each sample to contribute to both the pool and the sample to be hybridized. In addition, if the study is to be extended in the future with additional samples, it is best to use the same reference for all studies. The reference sample should always be either as closely related as possible to the samples in the study (but still containing all microRNAs likely to be found in the samples), or as complex as possible. A complex reference sample could be a pool of total RNA extracted from many different tissues and in which most microRNAs would likely be found.

A major advantage of the single channel data analysis is that it eliminates the need for a reference sample, and allows for easy cross experiment comparisons–provided that the inter-chip variation is low.

4.3. *Pre-processing of NCI-60 microarray data*

The variation of the expression measurements in microarray experiments originates from both biological differences between samples and experimental error. The technical variation can be introduced at many levels of the experiment; e.g. RNA extraction, labelling, hybridization and washing. The purpose of the pre-processing is to minimize the systematic technical variation, in order to expose the biological differences between RNA samples.

The pre-processing of microarray data is performed in three steps: a) background correction, b) normalization and c) averaging replicate spots to intensity of a probe. All pre-processing methods presented here are implemented in the R[25] package LIMMA[26], which is part of Bioconductor.[22]

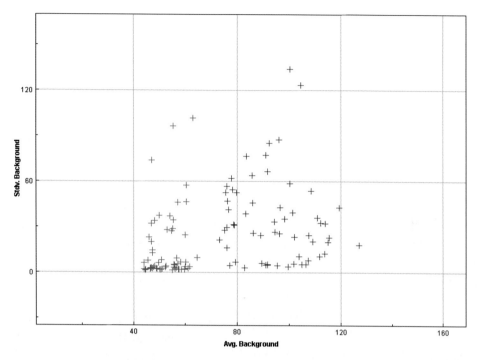

Figure 7. The standard deviation of the local background signals for each individual array plotted against the average local background signal. The red crosses are the common reference channel labeled with Hy5 and the green crosses are the individual cell line samples labeled with Hy3.

4.3.1. *Background correction*

The background correction is a process for adjusting spot intensities (foreground) for background fluorescence, which is caused by the combined effects of non-specific binding of labeled sample to the array surface and the optical noise of the scanner.

As seen from Figure 7 there is a difference between the intensities of the local background signal dependent on the channel. Simple background subtraction, i.e. subtracting local or global background from foreground is not recommended, as it often produces negative values, which are not amenable to logarithmic transformation. Another disadvantage of background subtraction is the increase in variation for low intensity spots. Several alternative background correction methods have been proposed, as reviewed by Ritchie *et al.*[27] Consistently with Ritchie *et al.*, we found the normexp + offset method to perform superior to the other methods in the paper and to local/global background subtraction. This conclusion was based on the analysis of diagnostic plots (MA plots). The normexp method is based on a convolution model, which ensures that the adjusted intensities are positive. The normexp method is explained in detail by Ritchie et al.[27]

The "normexp + offset" method is a modification in which a small positive offset is added to move the corrected intensities away from zero, resulting in variance stabilization of low intensity probes. We found that setting the offset to 10-20 is enough to eliminate all negative values.

4.3.2. *The normalization of two-color microarray data*

In two-color microarray experiments, the intensity of the Hy3™ channel (Green, G) is divided by the Hy5™ intensity (Red, R) for each spot, producing G/R ratios, which are a measure of relative expression. Because the labelling efficiencies and scanning properties of the two flourophores differ, this may cause transcript-dependent dye-bias, which leads to difference between the dyes e.g. the background levels of the two channels are not identical (see Figure 7). The dye-bias depends on spot intensity and can be visualized by means of an MA-plot, where $M = \log2 (G/R)$, and $A = (\log2 G + \log2 R)/2$. In other words, M and A represent the log2 transformed ratio and log2 of the geometrical mean between G and R, respectively (see Figure 8). In order to minimize the effect of the dye-bias on the log2 ratios (M-values), the global locally weighted scatterplot smoothing (loess) normalization is performed:

$$N = M - \text{loess}(A),$$

where loess(A) is the loess curve as a function of A, which is a robust smoother based on local polynomial regression. Loess normalization is applied to each array separately (within-array normalization) and it assumes that most genes targeted by probes on the array are not differentially expressed and therefore it centers the majority of M- values around zero (by subtracting the local, intensity dependent factor loess (A) from M-values)

As Figure 8 shows there is very little dye effect on the array, thus in principle one could use both channels for samples on the array. This will require somewhat extra care for setting up the experiments to exclude confounding variables such as hybridization and labelling between different days. The loess normalization can skew the data if the sample input (miRNA content) is very different from that of the reference channel. Furthermore, the method cannot be employed if one is interested in looking at global changes in miRNA expressions, as the underlying assumption is that the number of up- and down-regulated miRNAs is balanced between the samples. In such cases other methods have to be used to measure

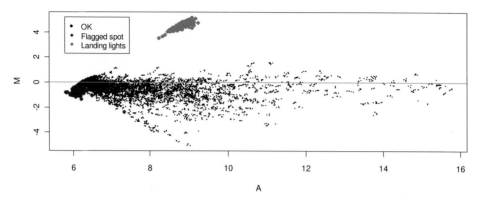

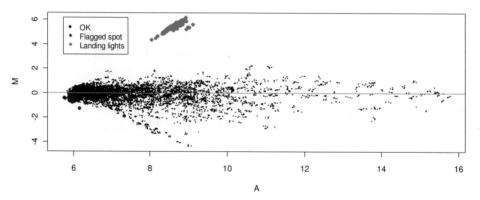

Fgure 8. MA plot illustrating the effect of loess normalization. In the plots the red spots are flagged features, which are the empty spots and technically failed spots, while the landing lights (grid alignment spots) are shown in green. The brain cell line SF-268 was selected as a representative example of how data responds to loess normalization. The top panel shows the expression data before normalization, while the second panel shows data after loess normalization.

and calibrate for differences in miRNA input. This could be done by purifying the small RNA fraction (< 30 nucleotides) and using this as input material for the labelling reactions. Another method employs addition of synthetic spike-in sequences to the samples for normalization.

The most commonly used methods for within-array normalization are described by Smyth and Speed[28] and include print-tip loess and global loess.

4.3.3. *Normalization of single-channel microarray data*

In the single-channel microarray experiments, the absolute probe intensities of Hy3 (Green channel) are used rather than ratios, and the normalization is performed across the arrays (between-array normalization) to enable direct comparison of spot intensities between arrays.

The normalization of single channel data is performed in two steps: first, the background corrected Hy3 intensities are log 2 transformed. Second, quantile normalization is applied, which forces the log2 intensities to have the same distribution across all arrays[29]. The assumption of quantile normalization is that the distributions of expression levels across samples are similar; in cases where the assumption cannot be met, other methods may be preferable e.g. MAD-scaling,[30] which ensures that the median average deviation (MAD) of spot intensities is the same across all arrays.

Rao *et al.* compared various methods for between channel normalization of miRNA microarray data.[31] The authors conclude that quantile normalization outperforms cyclic loess, mean and median normalization. The MAD-scaling was not addressed by the authors.

When conducting between array normalization, special attention should be paid to the number of not available values (NAs). First, the number of NAs should be kept as low as possible, which can be facilitated by choosing a background correction method, which avoids negative values. Second, setting the intensities of low intensity spots (so called "flooring") to NA or some artificial value should be avoided.

As seen in Figure 9 quantile normalization forces the intensities of capture probes to the same distribution with approximately the same median and spread. These plots, however, cannot show if the normalization of the data is appropriate for the particular data set. The raw data box plots have four times as many spots as the normalized data, because averaging of the spots is done after normalization. This averaging manifests itself as fewer outliers in the box plots.

Sarkar *et al.* have recently published a paper comparing different single channel normalization procedures for different miRNA array platforms.[23] They showed that batch effects (variance related to groups of samples labeled together) occur when conducting experiments across different days. Normalization is needed to remove these day to day variations; a good normalization procedure will remove the batch effect while keeping the biologically relevant difference between the samples. The authors did not test the quantile normalization method, but a couple of the other methods are discussed below.

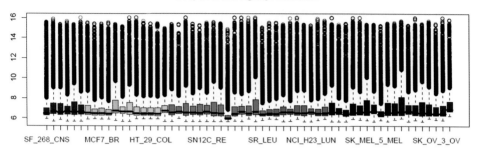

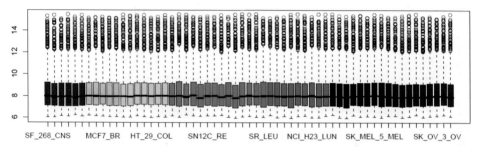

Figure 9. Box plots of log2 Hy3 sample channel data before (top) and after (bottom) quantile normalization. The boxes illustrate the median and the 25% (bottom) and 75% (top) quartiles. The outliers are values higher or lower than 1.5 times the interquartile range. Cell lines originating from the same tissue are represented by the same color; from left to right we have, red = brain (CNS), orange = breast, light green = colon, green = renal (kidney), light blue = leukemia, blue = lung, dark blue = melanoma, purple = ovary and pink = prostate.

As Sarkar *et al.* show, the use of global median gives a good and solid normalization of the probes in the higher intensity range, but performs poorly in the lower intensity end. This was shown by the presence of batch effects in the low intensity region. We therefore do not recommend the use of median scaling when also investigating variation in the low intensity region. Other methods used in the Sarkar *et al.* paper were VSN and VSN normalization with the use of spike-ins. Both methods performed well in their hands, but the choice of normalization method depends on the scope of the project. The addition of synthetic spike-ins could influence the normalization when using the spike-in normalization. This method requires accurate pipetting to ensure equal spike-in addition to all samples.

We have implemented the use of median average deviation (MAD) plots to illustrate the effectiveness of quantile normalization. To evaluate the effect of quantile normalization we compared the signals before and after normalization. We included the reference channel in the quantile normalization methods to see how this data, where no difference is expected, would behave when quantile normalized. As seen in Figure 10 the quantile normalization lowers the variance for the reference channel, as measured by the MAD-value across all probes for the unnormalized data, while preserving the variance in the sample channel. Only a few probes in the reference channel are above a MAD value of 0.4, which in our experience and compared to the values in the Sarkar *et al.* paper is a low variance for the reference channel. When comparing these values to the sample channel it is clear that much more variance is present—as expected—between the cell lines.

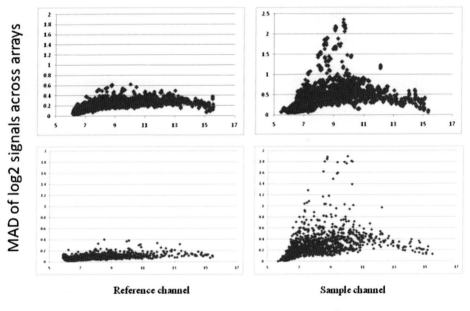

Median of log2 signal across arrays

Figure 10. The plots show the MAD of log2 signals across the arrays plotted on the y-axis against the median of log2 signals across the arrays on the x-axis. The top two panels are the unnormalized data from the reference channel and the sample channel, while the bottom two panels show the quantile normalized data.

4.3.4. *Day to Day Variation Using Single Channel Analysis*

Because the number of samples (59 samples) exceeded the capacity of our array hybridization station (12 slides per overnight run) it was necessary to run the arrays on different days. To estimate day to day variation, technical replicates of two cell lines were run as controls on each day. The reproducibility of technical replicates is visualized by pair-wise correlation plots of the quantile normalized log$_2$ intensities. The analysis is done based on quantile normalized data after averaging of replicate spots. The technical replicates of the two cell lines UACC-257 and HL-60 samples are highly correlated (correlation coefficient >0.99) and the pair wise correlation plots show very good agreement as only 1-2 outliers are present in the low intensity range. In contrast to the scatter plots of technical replicates, the scatter plots of UACC-257 against HL-60 samples show many spots with differential intensity, as expected (see Figure 11).

Such plots need visual inspection to check for reproducibility issues, as the correlation coefficients are high also between the different cell lines. This is due to the many spots in the low (unexpressed) region that give an artificially high

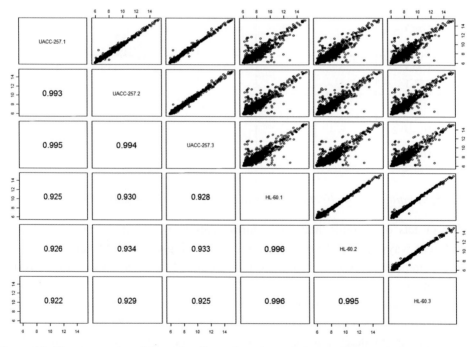

Figure 11. The scatter plots (above the diagonal) and correlation coefficients (below diagonal) between the probe log2 intensities of technical replicates. Technical replicates of UACC-257 and HL-60 samples were hybridized within three days (3 batches) on 21-23/Jan-2008. The log2 Hy3 intensities were quantile normalized. The plots show excellent correlation between technical replicates with only 1-2 outliers in the low intensity range. The scatter plots of technical replicates contrast with the plots of UACC-257 against HL-60 samples, where the differences are clearly visible.

correlation coefficient. Therefore, one can also look at how many miRNAs have higher fold change then a certain threshold, in this case: 2 (see Table 1).From Table 1 it is clear that many more miRNAs that are detected as differentially expressed between the different cell lines than between technical replicates. Only 5 probes detect more than 2-fold differences within pairs of the same sample run on different days, while—on average—approximately 120 miRNAs are more than 2-fold regulated between biological samples. This gives a false discovery rate of around 4% when using a 2-fold threshold as a cut-off for significance.

Table 1. Number of miRNAs with more than 2-fold difference between pairs of quantile normalized arrays run on different days.

	UACC-257.1	UACC-257.2	UACC-257.3	HL-60.1	HL-60.2	HL-60.3
UACC-257.1	0	6	4	124	128	126
UACC-257.2		0	5	105	109	101
UACC-257.3			0	121	119	121
HL-60.1				0	4	6
HL-60.2					0	4
HL-60.3						0

4.3.5. Evaluation of effect of different normalization methods

The effects of quantile and loess normalization of the data were evaluated in different ways. Correlation coefficients of the single channel intensity data and the common reference ratios were calculated and showed good overall agreement.

Two examples of correlations are illustrated in Figure 12, which shows good correlation between the two normalization methods.

In order to determine the impact of normalization method on high level analysis, we performed an analysis of differential expression across different tissue of origin based on both single and dual color analysis and compared the results. We looked for differences between cell lines of one tissue origin against the rest of the cell lines in the NCI-60 cell panel. Thus, miRNAs that are either up- or down-regulated in cell lines of each tissue compared to the rest were

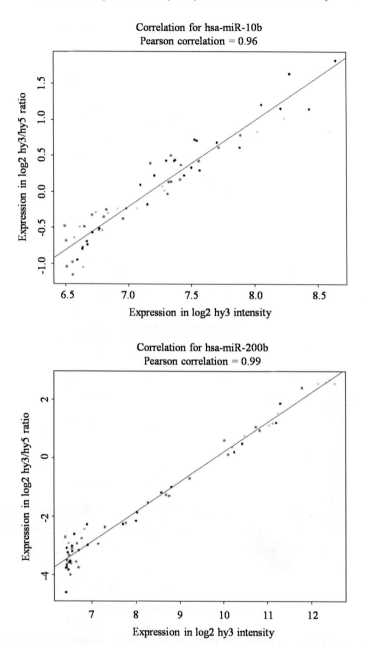

Figure 12. Scatter plots of single channel log2 intensities versus the log2 ratios. The upper plot represents high intensity probe (miR-10b) while the lower plot represents low intensity probe (miR-200b) data. Each point represents a measurement of a given miRNA in a cell line sample. Single channel intensities and ratios correlate very well in the case of high intensity signals. The good correlation of the low intensity probe indicates that single channel analysis can also be used reliably within the low intensity range.

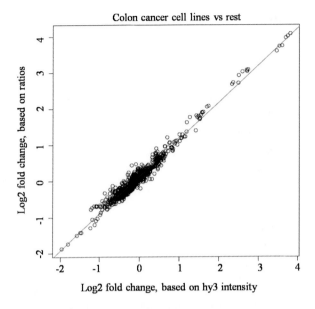
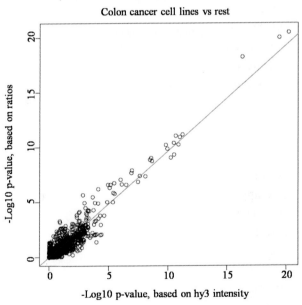

Figure 13. Comparison of results of differential expression analysis of miRNAs based on $\log_2$ ratios (y-axis) and $\log_2$ intensities (x-axis). In the analysis, a colon (cell lines originating from colon cancer) versus rest (all cell lines, excepting colon) was performed using both single channel and ratio data. Subsequently, the fold changes (left plot) and p-values (right plot) have been compared. The p-values were calculated by means of moderated t-statistics. The very good agreement between fold changes and p-values generated by the two pre-processing techniques indicate high reliability of the results.

identified. The significance was measured by means of moderated t-statistics, which is a modification of a simple Bayesian model to moderate standard errors across genes. In the results, the information from collection of miRNAs is used to improve the inference about each individual miRNA. The empirical Bayes, moderated t-statistic has been found to yield reliable results across a range of sample sizes,[32] therefore providing a robust method for general application. It is described in details by Smyth et al.[33]

Figure 13 shows an example of how the significance and fold change vary when looking at changes.

The good correlation of P-values between the two normalization methods suggests that similar results are obtained with both methods. Also, with respect to the fold change we obtain highly correlated results, so one would expect to detect the same difference with both methods. This also illustrates that the two normalization methods agree on the significantly differentially expressed miRNAs. Thus, the microarray data can be analyzed in both a single channel and a dual channel configuration, and the results from both analyses concur.

Clustering of the cell lines based on the most differentially expressed miRNAs identified by both the loess normalized ratios and $\log_2$ quantile normalized data gives similar results for the tightest clustering cell lines (Figure 14). Because clustering methods are very method sensitive by nature, the two clusterings are not expected to give exactly the same results.

5. Conclusion

In this chapter we have shown that LNA enhanced microarrays enable high quality miRNA expression profiling and that standard methods for array normalization can be applied. The quantile and the loess normalization methods for single and dual channel data respectively have been used for mRNA expression normalization for years and are implemented into both open source and commercial software packages. In our hands both methods were able to reduce the technical variation sufficiently to obtain highly concordant results for differentially expressed miRNAs. We do not recommend one method over the other, but suggest to use the method best suited for the intended purpose of the experiment at hand.

The good overall correlation and consistency of the NCI-60 dataset using both normalization methods illustrates the applicability of LNA enhanced arrays for single channel experiments.

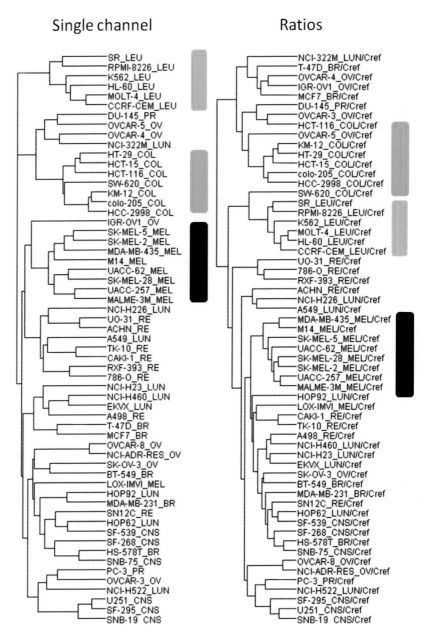

Figure 14. Unsupervised hierarchical clustering of cell lines based on the 100 most differentially expressed miRNA between the cell lines using Pearson correlation as distance measure for both data sets. The clusters of cell lines which show the most similarity are highlighted with boxes corresponding to the colors; light blue = leukemia, green = colon, and dark blue = melanoma.

References

1. Gregory RI, Yan KP, Amuthan G, Chendrimada T, Doratotaj B, Cooch N, et al. The Microprocessor complex mediates the genesis of microRNAs. Nature 2004 Nov 11; 432(7014):235-40.
2. Griffiths-Jones S, Saini HK, van DS, Enright AJ. miRBase: tools for microRNA genomics. Nucleic Acids Res 2008 Jan; 36(Database issue):D154-D158.
3. Lagos-Quintana M, Rauhut R, Lendeckel W, Tuschl T. Identification of novel genes coding for small expressed RNAs. Science 2001 Oct 26; 294(5543):853-8.
4. Bernstein E, Kim SY, Carmell MA, Murchison EP, Alcorn H, Li MZ, et al. Dicer is essential for mouse development. Nat Genet 2003 Nov; 35(3):215-7.
5. Calin GA, Dumitru CD, Shimizu M, Bichi R, Zupo S, Noch E, et al. Frequent deletions and down-regulation of micro- RNA genes miR15 and miR16 at 13q14 in chronic lymphocytic leukemia. Proc Natl Acad Sci U S A 2002 Nov 26; 99(24):15524-9.
6. Park CS, Tang SJ. Regulation of microRNA Expression by Induction of Bidirectional Synaptic Plasticity. J Mol Neurosci 2008 Nov 8.
7. Bicker S, Schratt G. microRNAs: tiny regulators of synapse function in development and disease. J Cell Mol Med 2008 Sep; 12(5A):1466-76.
8. Krichevsky AM, King KS, Donahue CP, Khrapko K, Kosik KS. A microRNA array reveals extensive regulation of microRNAs during brain development. RNA 2003 Oct; 9(10):1274-81.
9. Castoldi M, Schmidt S, Benes V, Noerholm M, Kulozik AE, Hentze MW, et al. A sensitive array for microRNA expression profiling (miChip) based on locked nucleic acids (LNA). RNA 2006 May; 12(5):913-20.
10. Vester B, Wengel J. LNA (locked nucleic acid): high-affinity targeting of complementary RNA and DNA. Biochemistry 2004 Oct 26; 43(42):13233-41.
11. You Y, Moreira BG, Behlke MA, Owczarzy R. Design of LNA probes that improve mismatch discrimination. Nucleic Acids Res 2006 May 2; 34(8):e60.
12. Landgraf P, Rusu M, Sheridan R, Sewer A, Iovino N, Aravin A, et al. A mammalian microRNA expression atlas based on small RNA library sequencing. Cell 2007 Jun 29; 129(7):1401-14.
13. Wu H, Neilson JR, Kumar P, Manocha M, Shankar P, Sharp PA, et al. miRNA profiling of naive, effector and memory CD8 T cells. PLoS ONE 2007 Oct 10; 2(10):e1020.
14. Morin RD, O'Connor MD, Griffith M, Kuchenbauer F, Delaney A, Prabhu AL, et al. Application of massively parallel sequencing to microRNA profiling and discovery in human embryonic stem cells. Genome Res 2008 Apr; 18(4):610-21.
15. Li J, Yang Z, Yu B, Liu J, Chen X. Methylation protects miRNAs and siRNAs from a 3'-end uridylation activity in Arabidopsis. Curr Biol 2005 Aug 23; 15(16):1501-7.
16. Neilson JR, Zheng GX, Burge CB, Sharp PA. Dynamic regulation of miRNA expression in ordered stages of cellular development. Genes Dev 2007 Mar 1; 21(5):578-89.
17. Ruby JG, Jan C, Player C, Axtell MJ, Lee W, Nusbaum C, et al. Large-scale sequencing reveals 21U-RNAs and additional microRNAs and endogenous siRNAs in C. elegans. Cell 2006 Dec 15; 127(6):1193-207.
18. Weinstein JN, Pommier Y. Transcriptomic analysis of the NCI-60 cancer cell lines. C R Biol 2003 Oct; 326(10-11):909-20.
19. Ikediobi ON, Davies H, Bignell G, Edkins S, Stevens C, O'Meara S, et al. Mutation analysis of 24 known cancer genes in the NCI-60 cell line set. Mol Cancer Ther 2006 Nov; 5(11):2606-12.

20. Blower PE, Verducci JS, Lin S, Zhou J, Chung JH, Dai Z, et al. MicroRNA expression profiles for the NCI-60 cancer cell panel. Mol Cancer Ther 2007 May 1; 6(5):1483-91.
21. Churchill GA. Fundamentals of experimental design for cDNA microarrays. Nat Genet 2002 Dec; 32 Suppl:490-5.:490-5.
22. Gentleman RC, Carey VJ, Bates DM, Bolstad B, Dettling M, Dudoit S, et al. Bioconductor: open software development for computational biology and bioinformatics. Genome Biol 2004; 5(10):R80.
23. Sarkar D, Parkin R, Wyman S, Bendoraite A, Sather C, Delrow J, et al. Quality Assessment and Data Analysis for microRNA Expression Arrays. Nucleic Acids Res 2008 Dec 22.
24. Hua YJ, Tu K, Tang ZY, Li YX, Xiao HS. Comparison of normalization methods with microRNA microarray. Genomics 2008 Aug; 92(2):122-8.
25. R Development Core Team. R: A Language and Environment for Statistical Computing. 2008.
26. Smyth GK, Speed T. Normalization of cDNA microarray data. Methods 2003 Dec;31(4):265-73.
27. Ritchie ME, Silver J, Oshlack A, Holmes M, Diyagama D, Holloway A, et al. A comparison of background correction methods for two-colour microarrays. Bioinformatics 2007 Oct 15; 23(20):2700-7.
28. Smyth GK, Speed T. Normalization of cDNA microarray data. Methods 2003 Dec; 31(4):265-73.
29. Bolstad BM, Irizarry RA, Astrand M, Speed TP. A comparison of normalization methods for high density oligonucleotide array data based on variance and bias. Bioinformatics 2003 Jan 22; 19(2):185-93.
30. Yang YH, Dudoit S, Luu P, Lin DM, Peng V, Ngai J, et al. Normalization for cDNA microarray data: a robust composite method addressing single and multiple slide systematic variation. Nucleic Acids Res 2002 Feb 15; 30(4):e15.
31. Rao Y, Lee Y, Jarjoura D, Ruppert AS, Liu CG, Hsu JC, et al. A comparison of normalization techniques for microRNA microarray data. Stat Appl Genet Mol Biol 2008; 7: Article22.
32. Jeffery IB, Higgins DG, Culhane AC. Comparison and evaluation of methods for generating differentially expressed gene lists from microarray data. BMC Bioinformatics 2006 Jul 26; 7:359.:359.
33. Smyth GK. Linear models and empirical bayes methods for assessing differential expression in microarray experiments. Stat Appl Genet Mol Biol 2004;3:Article3. Epub; 2004 Feb 12; Article3.

CHAPTER 3

ANALYSIS AND QC FOR REAL-TIME QPCR ARRAYS OF HUMAN MICRORNAS

Dirk P. Dittmer, Pauline Chugh, Dongmei Yang, Matthias Borowiak,
Ryan Ziemiecki and Andrea J. O'Hara

Dept. of Microbiology and Immunology, Lineberger Comprehensive Cancer Center, Center for AIDS Research. The University of North Carolina at Chapel Hill 715 Mary Ellen Jones CB#7290, University of North Carolina at Chapel Hill Chapel Hill, NC 27599-7290

MicroRNAs are a class of small, non-coding RNAs that can regulate the transcription and translation of many target genes. Recently, several QPCR assays have been developed to detect the expression of microRNAs at different stages of maturation. Here, we review these assays, data analysis and quality control in detail. Optimization of the limit of detection, sample size calculation and power analysis for these QPCR arrays will be discussed. Finally, we present a case study example on qPCR arrays for the expression of pre-microRNAs.

1. Introduction

Pre-miRNA profiling using real-time QPCR was developed by Schmittgen and colleagues[19,23,24,37,38] as well as our group.[29,43] Mature miRNA profiling has been the purview of commercial developers, most notably Applied Biosystems Inc. (ABI). ABI commercialized the TaqMan™ real-time QPCR assay for mRNAs and miRNAs. They developed the first real-time QPCR array for human miRNAs, which has been used successfully in cancer profiling (see[22,29] for examples). This assay relied on a miRNA-specific reverse transcriptase primer, which in its initial form necessitated a separate RT reaction for each miRNA to precede real-time QPCR reaction. This design was impractical for all but a few fully automated laboratories and extremely costly in terms of labor and supplies. Since then, a multiplexed reverse transcription kit has become available, which alleviates this problem, but introduces the many problems associated with multiplex PCR and RT-PCR. At present, the list price for the complete human miRNA real-time QPCR array is $75,000 for 150 reactions, or $500 per array, which mirrors the price for hybridization-based arrays.

Earlier we published an in depth description of general methods, laboratory set-up and the "wetlab" aspects of real-time QPCR arrays.[30] Not much has changed since then and not much differs in regard to general laboratory aspects of pre-miRNA quantification by real-time QPCR arrays from mRNA quantification. The chemistry of mature miRNA detection by real-time QPCR is different for different commercial assays. Our publication[10] may serve as a fairly complete case study in mRNA real-time QPCR array design and analysis.

At present, no commercial programs for the analysis of real-time QPCR arrays commercial or self-designed exist. However, this will no doubt change. The analysis of real-time QPCR array data is no different than the analysis of other microarray data and after a few real-time QPCR pre-processing steps any commercial, academic clustering or class discovery program can be used. The purpose of this chapter is to describe the individual steps for data cleanup, normalization and unsupervised clustering of real-time QPCR data.

2. MicroRNA Maturation and the Utility of Pre-miRNA Profiling

MicroRNAs (miRNAs) are a novel class of mammalian genes. They regulate the transcription and translation of many target proteins and have been implicated in normal development as well as carcinogenesis. Viruses also encode miRNAs. For example, in Kaposi's sarcoma-associated herpes virus (KSHV), all known miRs are grouped together in the viral latency region.[5,33,36] This organization is similar to mammalian miR gene organization where clustering has been observed for 50-70% of miR genes.[3] The maturation of miRs is the subject of active research (reviewed in[7]).

First, a pri-miR is transcribed by RNA polymerase II. It is capped and polyadenylated in the nucleus.[4] The pri-miR can be of any length and contain any number of clustered miRs. The pri-miR serves as substrate for the Drosha nuclease complex.[46] Drosha cleavage generates pre-miRs which serve as the precursor of one or two mature miRs. The pre-miRs reside in the nucleus and are ~70 nucleotides in length. The stability of the pre-miRs can vary.[33,37] In KSHV, the pre-miRs are stable since they can be detected by Northern hybridization. Overall, their levels correlate with the level of the mature KSHV miRs[5,33,36] (see reference[14] for an exception).

The pre-miRs are subsequently exported out of the nucleus with the help of Exportin 5 and serve as a substrate for Dicer in the cytoplasm. Mature miR levels can be regulated by modulating exportin-5 expression.[45] In the cytoplasm, Dicer processes the pre-miR into the mature miR and complementary strand, each comprising ~22nt in length. For some miRs, both the sense and anti-sense pre-miR strands serve as template for mature miRs.

The miRNAs have emerged as master regulators of cell lineage differentiation and key modulators of cancer (reviewed in[6]). At present the Sanger database has recorded 678 human miRNAs[16] each capable of targeting up to several hundred different mRNAs.

Mature miRNA profiling has previously been used to stratify lineage types and disease progression stages. Many tumor-specific and cell lineage-specific signatures have been compiled[21,42,44] and many others). Pre-miRNA profiling has also been used successfully to stratify human tumors.[19,23,24] We previously profiled pre- and mature miRNAs for PEL[29] in order to establish a PEL cancer signature. QPCR has been shown to be an effective form of miRNA profiling. Northern blotting has limitations including low throughput and poor sensitivity. Alternative high throughput profiling methods, like microarrays, require high concentrations of target input, show poor sensitivity for rare targets, a limited linear range and the need for post-array validation by real-time QPCR.

Therefore, QPCR appears to be a better method for a limited set of targets such as the ~650 human miRNAs, and it can be applied easily at the pre-miRNA level as well.

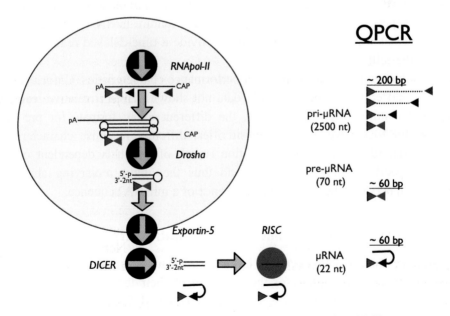

Figure 1. Outline of miRNA maturation and stage-specific real-time QPCR assays.

The miRNA genes are named according to the 60-80 bp sequence of the pre-miRNA segment.[15] Each miRNA gene locus produces one pre-miRNA, which in turn can produce one or two mature miRNAs depending on whether both strands of the mature product are functionally inserted into the RISC complex. While all miRNA genes, and therefore all pre-miRNAs, are made of unique sequence, different pre-miRNAs can be processed to yield an identical mature 22 nt miRNA. For instance, there are 3 different let-7a genes: let7-a-1, let-7a-2 and let-7a-3, each located on a different chromosome [9,13,22] and subject to different regulatory controls. Pre-miRNA profiling but not mature miRNA profiling distinguishes these genes.

How well do pre-miRNA levels correlate with mature miRNA levels? This seemingly simple question has a non-trivial answer.

(i) We and others have shown that pre-miRNA levels correlate well with mature miRNA levels.[19,23,29,37,38,43] However, there are well-documented instances, where SNPs can affect Dicer processing.[11,14] These exceptions are informative in their own right and only simultaneous quantification of pre- and mature miRNA levels can identify these.

(ii) The two assays (mature miRNA and pre-miRNA) measure two different events and thus provide non-redundant information. The pre-miRNA pool represents an intermediate step and thus responds without delay to changes in cellular transcription. Pre-miRNAs are co-transcriptionally processed.[4,27] They have a short half-life, much like mRNAs, and thus provide a sensitive read-out for the purpose of tumor profiling. By contrast, mature miRNAs are part of the relatively stable RISC complex and thus provide a time-delayed read-out of the state of the cell.

(iii) The two assays have different performance characteristics. Unfortunately, these are different for each miRNA (data not shown). Even if relative levels of pre- and mature miRNAs correlate, the different assay formats for pre- and mature miRNAs have different sensitivities, different response characteristics and a different lower limit of detection (much of which is dependent on the miRNA-specific primer sequences) and thus they have a varying ability to distinguish between the presence and absence of a miRNA sequence.

3. Primer Design for Real-Time QPCR Pre-miRNA Arrays

Polymerase chain reaction (PCR)[28] has allowed many scientific fields, including virology, to develop assays for the detection of their template of interest. In many instances, PCR has risen as the gold standard for the detection of the presence of a pathogen where cell culture or serological assays were once considered unsurpassed. However, post-PCR handling steps required to evaluate the product are a cumbersome part of PCR assays. The ability to track the amplification and quality of the product without post-PCR steps was first seen with the description of quantitative assays using replicable hybridization probes.[25] This technique has since become the foundation from which real-time quantitative PCR has been developed.[17] Real-time quantitative PCR measures the amount of PCR product at each cycle of the reaction either by binding of a fluorescent, double strand-specific dye (SYBRgreen™) or by hybridization to a third sequence-specific, dual-labeled fluorogenic oligonucleotide (molecular Beacon, TaqMan™). Since the introduction of real-time QPCR, many applications have arisen using this technology. Its kinetics and chemistries are covered in detail by Mackay *et al.*[26]

Primer design is one of the most important aspects in achieving a successful real-time QPCR assay. It is difficult to do for 22 nt long mature miRNAs, but possible for the 70 nt pre-miRNAs using standard programs. There are many computer programs and web based applications available to assist in the design of primers and probes. Of these, we rely on eprimer3 as provided by EMBOSS.

EMBOSS (European Molecular Biology Open Software Suite)[34] is a comprehensive collection of free open-source programs for sequence analysis. It represents a freely available and more robust alternative to proprietary programs such as PrimerExpress (Applied Biosystems Inc., CA) and others.

Eprimer3, a program for searching PCR primers, is based on the Primer3 program[35] from the Whitehead Institute/MIT Center for Genome Research. It allows one to search a DNA Sequence for both PCR primers and oligonucleotide beacons. More than 60 parameters can be specified to adapt the program for various purposes. They include constraints on physico-chemical properties of the primers, probes and product, like TM, GC content and size; constraints on sequence properties, like the amount of self-complementarity and 3'-overlapping bases; positional constraints within the template sequence; avoidance of sequences specified in a mis-priming library, and many more. Detailed examples for use of Eprimer3 can be found in.[30] We have developed the following guidelines for pre-miRNA primer design:

i. The melting temperature (Tm) of the primers should be in the range of 59±2°C.
ii. The maximal difference between two primers within the same primer pair should be less than or equal to 2°C.
iii. Total Guanidine (G) and Cytosine (C) content within any given primer should be between 20-80%.
iv. There should not be any GC clamps designed into any of the primers.

v. Primer length should fall into the range of 9-20 nucleotides.
vi. Hairpins with a stem length greater than or equal to 4 residues should not exist in the primer sequence.
vii. Fewer than four repeated G residues should be present within a primer.
viii. The resulting amplicon should be at least 40 nucleotides in length.

4. Power Analysis and Sample Size Calculation for dCT

The overarching goal of this chapter is to provide bioinformatics approaches and tools for the analysis of real-time QPCR arrays that are as comprehensive as needed, but not more complicated than they should be. The limit in real-time QPCR array analysis is biological variation. We can never expect to go beyond it. Even the most sophisticated algorithms cannot make up for the sample-to-sample variation inherent to biological processes. This is captured by power analysis and sample size calculations.

For instance, to determine what range of responses can be expected, we analyzed mRNA transcription in two Burkitt's lymphoma (BL) cell lines (one being sensitive to the drug AZT, one being resistant). Based upon hierarchical clustering, the most changed mRNA coded for vBCL (an anti-apoptosis gene). We analyzed n=13 independent samples taken over a 24 hour time course and normalized total mRNA levels to HPRT (a "housekeeping gene"). We set the mean for one cell line to zero (ddCT normalization[17]) to obtain relative changes and calculated fold differences as fold = 2^{-ddCT} (Table 1, note that increases in mRNA levels appears as negative ddCT). Since all EBV+ Burkitt's lymphomas require the gene EBNA-1 to maintain the viral episome, EBNA-1 levels did not change significantly ($p \geq 0.1$). This mRNA therefore functions as a true experimental null hypothesis. By comparison, the vBCL-2 mRNA was increased $\geq$170 fold in the AZT resistant cell line compared to the AZT sensitive cell line ($p \leq 0.005$).

Table 1. Observed change in mRNA levels and sample variation levels in cell lines.

genes	AZT sensitive BL ddCT (n=13)		AZT resistant BL ddCT (n=13)		fold 95%CI	p(t-test)
	µ1(ddCT)	SD	µ2(ddCT)	SD		
vBCL-2	0	0.55	-7.74	0.57	(173 to 266 x)	$\leq 1 \times 10^{-16}$
EBNA-1	0	0.61	-0.5	0.59	(0.6 to 1.76 x)	≥ 0.1

As we expect more sampling error in primary tumor biopsies as compared to cell lines, we performed a similar calculation using our data on primary Kaposi sarcoma biopsies.[9] Firstly; we examined mRNA levels for LANA/orf73 mRNA, which is required to maintain the tumor. As expected, we showed that LANA mRNA was present in every tumor cell by in situ hybridization.[8] This mRNA therefore functions as a true experimental null hypothesis. Secondly, we examined mRNA levels for the vGPCR gene, which based on our studies[9] and in situ studies by others[20] varies considerably in these tumors. The tumor samples were divided into two groups by unsupervised clustering and mRNA levels compared by t-test. As before, the CT values were normalized to a "housekeeping gene" to adjust for total RNA concentration and fold differences were calculated based on fold = 2^{-ddCT} (Table 2).

Table 2. Observed changes in mRNA levels and sample variation in biopsies.

genes	KS tightly latent ddCT (n=11)		KS with lytic foci ddCT (n=10)		fold 95%CI	p(t-test)
	µ1(ddCT)	SD	µ2(ddCT)	SD		
vGPCR	0	3.1	-9.8	3.6	(223 to 3326 x)	$\leq 1 \times 10^5$
LANA-1	0	2.6	-1.9	3.7	(0.8 to 12 x)	≥ 0.1

Since all KSHV+ KS require LANA to maintain the viral episome, LANA levels did not change significantly between groups ($p \geq 0.1$). By comparison, KSHV vGPCR was induced ≥ 223 fold in approximately half of all KS samples ($p \leq 0.005$), which suggested that these tumors contained a greater fraction of lytically reactivating cells and will be susceptible to anti-viral drugs. As expected, the variation (SD) is greater in clinical biopsies compared to clonal cell lines since the samples represent different stages of tumor development.

Based on the effect size, we can calculate the sample size that is required to conclude a difference in mean ddCT (two-sided), type-II error of 85% accuracy (Figure 2). For the purpose of power analysis, we use the more stringent alpha ≤ 0.005 to account for multiple comparison testing rather than the customary alpha ≤ 0.05. Therefore, based upon our QPCR accuracy, less than 20 biological replicates are needed to detect a 10-fold difference in mRNA levels between treatments.

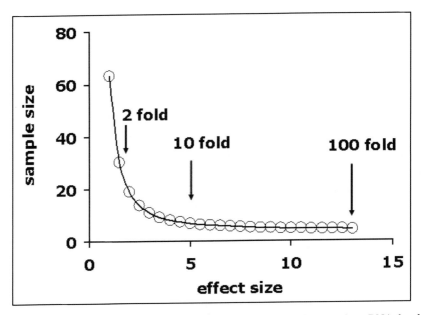

Figure 2. Number of samples/replicates required to conclude a difference in mRNA levels by QPCR. Effect size is given in CT units on the horizontal axis and sample size on the vertical axis.

5. Calculating Biologically Relevant Mean CT for Replicate Measurements

The mean is a derivative measure of central tendency. The analysis of central tendency also provides measures of spread, such as the standard deviation (SD). For instance, in an experiment using technical triplicate measurements we replace the raw individual CTs for gene A with the mean CT (meanCTA±SD). Reporting a mean (or median) without the SD leaves out essential information that helps the reader assess the quality of the measurement. For real-time QPCR data, the SD is reported in CT units. A large amount of vibrant literature exists regarding error calculations for real-time QPCR measurements.[31,32,41] Not reporting the SD or a comparable measure in variance could mean rejection of the final manuscript.

The SD does not take into account the number of replicates and is dependent on the absolute value of mean CT. Hence, it is almost always advisable to report the standard error of the mean (S.E.M.) instead, which is calculated as S.E.M. = SD/SQRT (number of replicates). The 95% confidence intervals (C.I.) are even more informative and are easily calculated in any spreadsheet or statistical program.

We sometimes use a shortcut with regard to real-time QPCR array analysis, which is designed to give ONE central measure of variability across all samples and all primers for a given data-set.

i. Calculate the SD for each primer across all samples and replicates.
ii. Calculate %SD as SD/median for each primer across all samples and replicates.
iii. Report the five number summary of the %SD; i.e. the smallest observation, lower quartile (Q1), median (Q2), upper quartile (Q3), and largest observation.

In this case, we do not report individual C.I. for each primer. One would expect useful biomarkers for the data set in question to be in the upper quartile and useful normalizing genes to be in the lower quartile. This was previously introduced for conventional microarrays.[39] Except cluster analysis then uses this information as a selection tool: only Q3 is used for subsequent cluster analysis, since any variation < Q3 can be considered insignificant. Also, this analysis identifies outliers for manual inspection, which could result from data entry errors, primer failures or very differentially expressed miRNAs.

There is an alternative philosophy with regard to real-time QPCR array analysis, which we use for unsupervised clustering of large data sets. We use all raw CT data, normalize by median of array (see below for details) and subject this data to Euclidian-metric based clustering and heat-map display. This highlights outliers of repeat measurements as well as outlier samples visually. This easily identifies outlier measurements for imputation, and sometimes this represents the most useful tool for miRNA biomarker discovery efforts. Since no data are filtered out, using the whole data set and nothing but the whole data set most truthfully resembles the variation of the experiment. Using all the raw data yields to the least biased determination of false discovery rate[40] for the data set.

There are many means, which by no means are the same: (i) the arithmetic mean or average of all data, (ii) the median of all data and (iii) percentiled mean or median of a fraction, typically 95%, of the data.

There are other means, still, but they are not relevant to real-time QPCR array analysis since CT data are log-transformed and thus can be averaged additively rather than geometrically as fractions would be.

The mean is only meaningful for $n \geq 3$ samples. Even though we can calculate the average of a pair (n = 2) of data points, we cannot compute the mean of duplicates. Hence, technical and biological repeats should be done in at least triplicates; better yet in quadruplicates (n = 4). Quadruplicates are convenient technically, because four replicates of 96 samples fit in a 384 well plate and can be pipetted quickly using a 96-tip pipetting head, such as a Hydra™ (Thermo Inc.). The more replicates, the smaller the variation, which allows us to distinguish smaller differences with accuracy. As a rule of thumb, we sometimes use sextuples to quantify 1.5 fold differences between samples.

The median is only meaningful for an uneven number of samples, $mod(n/2) = 0$, since it is a ranking based measure. Nevertheless, many programs such as Microsoft Excel will report a median for an even number of samples. The value is calculated as the average of the two data points that bracket the median. This is acceptable for everyday use.

The mean and median are only meaningful if the individual CT data are within the linear range of the assay. Hence, the linear range for each primer in the real-time QPCR array needs to be determined at least once. The problem typically is not signal saturation, because a given miRNA or pre-miRNA is too abundant. Rather, the difficulties arise at the limit of detection where even real-time QPCR is no longer linear, and where the maximum cycle number maxCT introduces an artificial threshold.

6. Real-Time QPCR Data Analysis at the Limit of Detection

All possible probabilities are distributed between 0 and 1. Most simple statistical analyses assume that the data follow a normal distribution. The normal distribution extends from minus infinity to plus infinity and is symmetrical around the mean. Real-time QPCR data are not because they are bounded by the maximal cycle number maxCT. It used to be that any PCR signal that required more than 35 cycles was considered contamination. Real-time QPCR pushed the limit to 40 cycles and now using automation and extra precautions, we can now run 50 cycles without accumulating any signal or product in the non-template control (NTC) reaction.

From a theoretical perspective, all real-time QPCR reactions should be run until a signal is detected in the NTC reaction. Only this, rather than any arbitrary cut-off determines the true background. In fact, the Roche LC480 software allows the user to extend the number of cycles for any given run in increments of 10. Since primer-dimer extension accounts for the majority of background signal, the true background determined by iterative extension of the cycle limit will be different for different primer pairs. This poses a variety of different problems.

Practical real-time QPCR arrays use a single fixed maxCT for every primer in the array for every run. How do we incorporate this singularity into the statistical analysis of real-time QPCR arrays?

i. We can simply ignore it and treat maxCT equivalent to any other CTA, with CTA < maxCT. This works surprisingly well on decent data sets. It has some unwelcome consequences when we attempt to calculate medians of multiple measurements or when we attempt to calculate statistical measures of significance.
ii. We can omit all maxCT from the analysis and treat the data as not available (NA). This is a purists approach that will yield a statistically sound and congruent analysis. Depending on the data set, it may force us to leave out a major portion of the data and thus degrade array performance.
iii. We can randomly replace maxCT with either maxCT+SD or maxCT-SD in the entire data set. This allows us to calculate a pseudo SD for those replicate measurements where all individual data points equal maxCT, i.e. where there is no variation at all. Then we can conduct a statistical analysis using standard measures.

Figure 3 exemplifies this problem. We compared two primer pairs A and B against the same target in a checkerboard design, i.e. with alternating positive and water wells. We used n = 96 repeats for each condition yielding a total of 384 QPCR reactions. We ran the QPCR for 55 cycles on a LC480 light cycler. Using primer A on our water control, we find that most reactions did not yield any signal (Figure 3, left panel). Hence, the spike in density at CT = maxCT = 55, representing 69 of 96 repeats.

Theoretically, all reactions that yield CT = maxCT could represent machine or pipetting errors. This is unlikely ($p \leq 2.2*10-16$ by X^2), since one would expect an equal number of such errors in the positive control, where there are none (0/96). This justifies counting all maxCT as CT for subsequent calculations rather than excluding them and assigning "NA", not available. This argument is highly dependent on the number of repeats.

Suppose we only performed 6 replicates and had obtained similar proportions of maxCT for water (4/6) and positive control (0/6). Here, $p > 0.05$ by X^2 and we could therefore not conclude that any of the maxCT were not pipetting errors and we have to assign "NA". If all water control reactions (6/6) had yielded maxCT, we could again use maxCT, since $p \leq 0.004$. It is essential that 100% of the water or NTC controls yield maxCT.

Using primer B on our water control, we find that almost half of the reactions did yield a signal (Figure 3, right panel). The fraction of maxCT = 55 was 58 of 96 repeats. Still $p \leq 2.2*10-16$ by X^2, since we had no reaction failures in the positive control reaction.

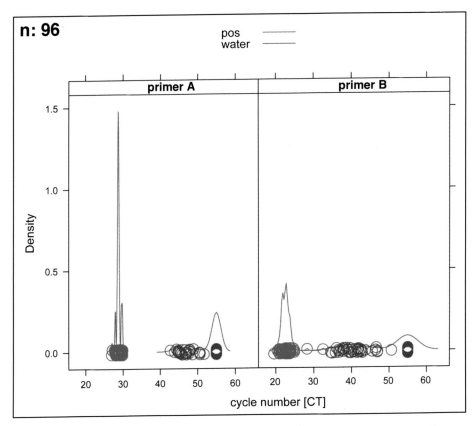

Figure 3. Density distribution of QPCR results for two primer pairs directed against the same target. Density is plotted on the vertical and cycle number (CT) on the horizontal axis. N: 96 replicates for positive (pos) or water control.

Primer B seems more sensitive, since for the same amount of input the medianCT = 22.64 for the positive reaction compared to primer B with medianCT = 28.69. However, CTs for the water control, i.e. false positives, also come up as early as 34.5 cycles. Using a conservative approach, the signal to background ratio is medianCT − 34.5 = 11.86 or ~ 3,700 fold. For primer A the ratio is medianCT − 42.7 = 14.01 or ~ 16,500 fold! Hence, even though primer A is less sensitive, overall it will perform better.

Another difference is conformity. Primer B had two outliers in the positive control with CTs of 28.21 and 32.41 respectively. This is not a problem with 96 replicates since they represent 2/96 < 2% of the data. Using either the median or the 10% trimmed mean would remove these outliers. We encountered a similar problem comparing different commercial real-time QPCR mixes[18]: here, too, the brighter mix was more sensitive in yielding a signal at earlier CT absolute terms. However, the NTC reactions also came up earlier and the variation was larger.

In sum, pipetting accuracy, contamination control and propensity for primer dimer formation determine the maximal cycle number for any real-time QPCR array. In most cases running real-time QPCR arrays for > 40 cycles is meaningless. A conservative cut-off is better, as it lowers the false positive rate. To statistically improve assay accuracy for low abundance miRNAs, more replicate arrays (~ 6) and more NTC reactions should be performed.

7. CASE STUDY: Analysis of Real-Time QPCR Arrays for Pre-miRNAs

For real-time QPCR arrays, types of normalization can be applied. Type I normalization relative to a reference sample t0 yields dCT. Type I normalization is applied for each target/primer pair. It can also be called "normalization by row", since in traditional microarray experiments the primers/genes are organized in rows and the samples in columns.

If there is no designated reference sample we can also calculate dCT based on the median CT for the target in question. This is analogous to median centering by gene as introduced by Eisen et al.[12] for microarrays. Type II normalization is relative to the reference gene e.g. U6 RNA. This eliminates differences due to variation of the overall input cDNA concentration. One should always aim to set up and experiment and normalize the input material (e.g. number of cells or total RNA) such that the variation in the reference gene is ±1 x CT unit.

During type I normalization, only CT values of a single primer pair are compared to each other. Hence, amplification of efficient differences between primer pairs do not enter the calculation.

In contrast, type II normalization compares two different primer pairs, such as for miRNA A and U6, with associated, possibly different, amplification efficiencies kA and kU6. This is a serious problem in real-time QPCR analysis. However, we can ignore this problem since for array analysis, clustering is performed in log-space (CT values) rather than interpolated RNA levels. Hence, primer efficiency does not play a role as long as all primers are reasonably similar ($1.8 < $ mean$K_{eff} < 2$) and only a linear term is subtracted during normalization, which does not impact the rank order between samples. Primer pairs with highly aberrant K_{eff} should be excluded from cluster analysis and analyzed on an individual basis.

Figure 4 exemplifies this result. Here, pre-miRNA was isolated as per our prior procedures[29] and miRNA levels quantified using SYBR-based real-time

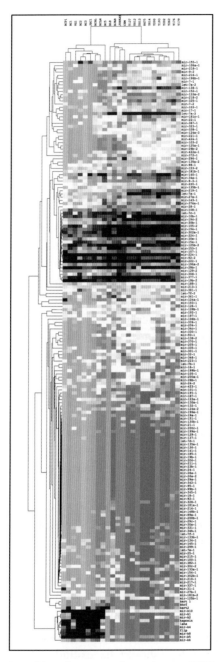

Figure 4. Two-way clustering of lymphoma samples and pre-miRNAs. High levels are shown in red, absent levels in blue and low, but detectable levels in white. The lower left corner shows pre-miRNAs associated with primary effusion lymphoma. The 7 rightmost samples are 2x2 mm tonsil biopsies (from our publications[13, 29] and unpublished).

QPCR. Data were normalized to U6 RNA as described above and clustered using Arrayminer™ (Optimal Design Inc., Nivelles, Belgium). We used a standard correlation metric, which does retain a measure of relative levels across the entire array.

The miRNAs, which show predominantly blue colors (top), are not expressed to any appreciable level compared to miRNAs, which show predominantly red colors across all columns. The most informative miRNAs are those in the middle of the heatmap. They show a large variation (blue to red) and split the sample columns, neatly into two groups.

Alternatively, one could use a Euclidian metric for clustering. However, the Euclidian metric exaggerates differences, to the point that only the most abundant miRNAs show variation in the heatmap. The Euclidian metric is not sensitive enough in most instances.

There is also the option of using a Pearson correlation coefficient based clustering, which in Arrayminer is coupled to ±1 normalization, i.e. the dCT for each primer pair are adjusted further such that the range is within ±1 of the median for that particular primer pair. Here, the median for each primer pair is artificially set to 0. Information of both the relative abundance of miRNAs to each other, as well as of the magnitude of the relative change is lost. This method makes for pretty pictures, but loses information that is essential to assess biological significance.

For instance, a more highly expressed miRNA is easier to detect, more likely to be within the middle of the linear range than on the fringes and is thus a better biomarker. In Pearson correlation coefficient or rank-based clustering, it is impossible to discern such a "good" biomarker, from a miRNA, which is only marginally present at all or from one, which shows only marginal variation. As demonstrated above, targets that show marginal variation, regardless of abundance, require many more samples to reach statistical significance.

In sum, real-time QPCR is ideally suited for miRNA and pre-miRNA profiling, since the total number of human miRNAs is around ~ 700 or two 384 well QPCR plates. In order to use real-time QPCR arrays for profiling, automation is a must, since the degree of replicate variability and the total number of replicates per sample correlate with the degree of statistical significance that can be attributed to a change in miRNA levels (CT). Real-time QPCR array data can be analyzed by the same statistical and bioinformatics methods as hybridization-based microarray data. There is an added benefit, though. If all primer pairs within the array perform with similar efficiency, relative fold changes of any one miRNA between samples and amongst different miRNAs across samples can be calculated as 1.8^{ddCT}.

Appendix: Primer pair verification with BLAST and EMBOSS

a. WEB-based NCBI BLAST

Many real-time QPCR primers are published by others or available from collections. These should be used whenever possible, since it allows for easy comparison between new and existing studies. However, they should be thoroughly verified by bioinformatics. Here, we will discuss some scripts for verifying existing primers through bioinformatics approaches.

First, all primers should be loaded into an Excel spreadsheet in a uniform format such as exemplified below. There are a number of nomenclature issues to consider. All primer sequences are 5' to 3' direction. All primer sequences should be in lower case. If necessary, use the Excel function **LOWER** () to convert each entry. The table should be in a fixed space font such as Monaco or Courier. This will enable visual alignment. Forward and reverse primers should be indicated by a common denominator such as "f" and "r". The names should contain no blanks (use underscore "_" if necessary and no special, non-ASCII characters. If known, a Genbank ID to the target sequence should be provided as well as a description, but this is optional as the target can be identified by blast search.

Table 3. Primer table in EXCEL.

description	genbankID	name	forward primer	name	Reverse primer
alpha-1-antitrypsin	nm_000295	000295f	tttagaggccatacccatgtc	000295r	ccactttcccatgaagagg
angiotensinogen	nm_000029	000029f	ttgagcaatgaccgcatc	000029r	ttgttaagctgttgggtagactc
apolipoprotein C-III (APOC3)	nm_000040	000040f	cagttccctgaaagactactgg	000040r	acggctgaagttggtctga

To confirm primer identity manually, use NCBI blast at http://www.ncbi.nlm.nih.gov/BLAST/. This program provides many options, but the easiest is to use the default blast program against the database "nr". nr contains all sequences in Genbank including mouse, human, animals, bacteria and viruses. Specifically, nr combines All GenBank+RefSeq Nucleotides+EMBL+DDBJ+PDB sequences (but no EST, STS, GSS, or phase 0, 1 or 2 HTGS sequences). As of 2007 it is no longer "non-redundant".

It is important to first concatemerize the forward and reverse primer using the Excel function **CONCATENATE** ().

This yields cagttccctgaaagactactggacggctgaagttggtctga for entry one. Then paste into the blast web interface and hit BLAST. BLAST will automatically adjust its parameters to optimize your search. See[1,2] for an in depth discussion. Unfortunately, BLAST will give many, many alignments. These can be cropped and formatted using parameters but most times the best alignment is listed first as in our example. Shown in Appendix Figure 1 is an example BLAST output. The right alignment will show 100% identity and 0% gaps for each primer. It will show both primers aligned to the same genbank entry, which in our example is: **ref|NM_000040.1|**

This entry points to a reference gene, which rather than any other entry should be reported. Right below the sequence identifier is the length of the corresponding genbank entry. Subtracting the start of the first primer, here 247, from the start of the second primer, here 336, gives the length of the amplicon, 89 base pairs in our example. Importantly, the first primer should align in orientation Plus/Plus and the second primer in orientation Plus/Minus or vice versa. If both primers align in the same orientation, they will not yield a PCR product.

In case of RT-PCR, the Plus/Minus primer can be used for specific priming of the RT reaction, since genbank records the sense strand for reference mRNAs. This may not be true for viral sequences or DNA targets.

If more than a handful of primer sequences are to be mapped onto targets, further automation is useful. We use the EMBOSS program. This requires the UNIX operating system with bash shell such as available on Linux or MacOsX based computers. It also requires a local installation of EMBOSS.

b. EMBOSS-based FUZZNUC

Under EMBOSS we can use the **fuzznuc** command for single primer match. This command takes an input sequence, here **NC_007605.fasta** and finds all occurrences of a pattern, here **AAATGGGTGGCTAACCCCTACATAA**. The number of mismatches can be specified as well as an output file, here **outfile**. Unlike **BLAST**, **fuzznuc** only searches a single entry in a single strand direction. Hence, fasta-formatted files can be used directly. Note, however, that the complement option in **fuzznuc** searches the complement, not the reverse complement strand.

To apply **fuzznuc** to the reverse complement strand, the EMBOSS command **revseq** must be run first.

Terminal 1 (operator input in bold):

BigMac-2:~/orf50Emboss dirk$ **fuzznuc NC_007605.fasta -pattern AAATGGGTGGCTAACCCCTACATAA -pmismatch 4 outfile**
Nucleic acid pattern search

BigMac-2:~/orf50Emboss dirk$ **fuzznuc NC_007605.fasta -pattern AAATGGGTGGCTAACCCCTACATAA -pmismatch 4 pan.txt -complement**
Nucleic acid pattern search

c. EMBOSS-based BL2SEQ

Under EMBOSS, we can use the **bl2seq** command for single primer match. This program has several parameters: -p blastn is the name of the program to be used, here blastn; -i specifies the first sequence in fasta format; -j specifies the second sequence in fasta format; -o denotes the output sequence and –D the output format. Terminal 2 exemplifies **bl2seq** using the primer **orf57f1** as input, **NC_009333kshvF.fasta** as target sequence and **test1** as output.

By feeding the output into **grep** with an arbitrary cutoff as defined by the pattern "e-" gives only the perfect match.

Terminal 2 (operator input in bold):

BigMac-2:~/KSHVprimers dirk$ **bl2seq -p blastn -D 1 -i orf57f1 -j NC_009333kshvF.fasta -o test1**

BigMac-2:~/KSHVprimers dirk$ **cat test1**
BLASTN 2.2.15 [Oct-15-2006]
Query: orf57f1
Fields: Query id, Subject id, % identity, alignment length, mismatches, gap openings, q. start, q. end, s. start, s. end, e-value, bit score
orf57f1   gi|139472801|ref|NC_009333.1| 100.00   23   0   0   1   23   82191   82213   2e-08   46.1
orf57f1   gi|139472801|ref|NC_009333.1| 100.00   11   0   0   12   22   21755   21745   0.35   22.3
orf57f1   gi|139472801|ref|NC_009333.1| 100.00   11   0   0   8   18   95339   95349   0.35   22.3

BigMac-2:~/KSHVprimers dirk$ **bl2seq -p blastn -D 1 -i orf57f1 -j NC_009333kshvF.fasta | grep -i e-**
Fields: Query id, Subject id, % identity, alignment length, mismatches, gap openings, q. start, q. end, s. start, s. end, e-value, bit score
orf57f1   gi|139472801|ref|NC_009333.1| 100.00   23   0   0   1   23   82191   82213   2e-08   46.1

d. SHELL-based manipulation of a real-time QPCR primer list

To ready an arbitrary long list of primers such as an entire 96 well plate for analysis first, using **CONCATENATE()** again, add "prim" in front of each primer without any spaces. Convert all Excel entries to a 2-column tab delineated text file. The file should be saved as **primers.text** and should look like indicated below. Since UNIX file designations are case sensitive, but those of other operating systems are not necessarily, it pays to only use lowercase letters for filenames. Again, there should be no spaces or special characters in the file name. Using a free text editor such as Textwrangler, open the file and remove all characters that don't pertain to the primers (from http://www.barebones.com/products/textwrangler/index.shtml).

Next, use the program to replace all the tabs with a single white space. After visual inspection, save the file again using the option "UNIX line breaks".

Terminal 3: The file: **primers.txt**

```
prim000295r    ccactttcccatgaagagg
prim000029r    ttgttaagctgttgggtagactc
prim000040r    acggctgaagttggtctga
prim000295f    tttagaggccatacccatgtc
prim000029f    ttgagcaatgaccgcatc
prim000040f    cagttccctgaaagactactgg
```

Now we can UNIX and EMBOSS utilities to convert each primer to a fasta formatted file. First, we use **sed** to introduce the ">" character as shown in terminal 1. Our Excel text file here is called **primers.txt** and our output file **400array**. You should visually inspect the output file, since **sed** will also put ">" in front of prim if a primer name contains prim somewhere in the middle.

Terminal 4 (operator input in bold):

BigMac-2:~/KSHVprimers dirk$ **sed -n 's/prim/>prim/p' primers.txt >400array**
BigMac-2:~/KSHVprimers dirk$ **less 400array**
BigMac-2:~/KSHVprimers dirk$ **head -n 2 400array**
>prim000295r ccactttcccatgaagagg
>prim000029r ttgttaagctgttgggtagactc

Next we introduce line breaks using a little script called awkscript.sh (or you can type **awk '$2 {$1 "\n" print $2}' 400array > array2** at the command line) as shown in Terminal 5.

Terminal 5 (The script awkscript.sh):

```
#!/bin/awk -f
# this part names the script
# Don't forget to use: chmod 755 awkscript.sh to make this file executable.
# You start the script with: ./awkscript.sh
{print $1 "\n" $2}
```

Once saved, we proceed as shown in Terminal 6:

Terminal 6 (operator input in bold):

BigMac-2:~/KSHVprimers dirk$ awk -f ./awkscript.sh 400array > 400array2
BigMac-2:~/KSHVprimers dirk$ **head -n 2 400array2**
>prim000295r
ccacttttcccatgaagagg
>prim000029r
ttgttaagctgttgggtagactc

The output file **400array2** is now in fasta format and can be used in any sequence analysis program, commercial or home-made. It can also be uploaded to web-based NCBI BLAST.

```
>ref|NM_000040.1|  Homo sapiens apolipoprotein C-III (APOC3), mRNA
Length=533
Sort alignments for this subject sequence by:
 E value   Score   Percent identity
 Query start position   Subject start position
 Score = 46.1 bits (23),   Expect = 0.002
 Identities = 23/23 (100%), Gaps = 0/23 (0%)
 Strand=Plus/Plus

Query  1    CAGTTCCCTGAAAGACTACTGGA  23
            |||||||||||||||||||||||
Sbjct  247  CAGTTCCCTGAAAGACTACTGGA  269

 Score = 38.2 bits (19),   Expect = 0.43
 Identities = 19/19 (100%), Gaps = 0/19 (0%)
 Strand=Plus/Minus

Query  23   ACGGCTGAAGTTGGTCTGA  41
            |||||||||||||||||||
Sbjct  336  ACGGCTGAAGTTGGTCTGA  318
```

Appendix Figure 1. Example of BLAST output

References

1. Altschul, S. F., and E. V. Koonin. 1998. Iterated profile searches with PSI-BLAST--a tool for discovery in protein databases. Trends Biochem Sci 23:444-7.
2. Altschul, S. F., T. L. Madden, A. A. Schaffer, J. Zhang, Z. Zhang, W. Miller, and D. J. Lipman. 1997. Gapped BLAST and PSI-BLAST: a new generation of protein database search programs. Nucleic Acids Res 25:3389-402.
3. Altuvia, Y., P. Landgraf, G. Lithwick, N. Elefant, S. Pfeffer, A. Aravin, M. J. Brownstein, T. Tuschl, and H. Margalit. 2005. Clustering and conservation patterns of human microRNAs. Nucleic Acids Res 33:2697-706.
4. Cai, X., C. H. Hagedorn, and B. R. Cullen. 2004. Human microRNAs are processed from capped, polyadenylated transcripts that can also function as mRNAs. Rna 10:1957-66.
5. Cai, X., S. Lu, Z. Zhang, C. M. Gonzalez, B. Damania, and B. R. Cullen. 2005. Kaposi's sarcoma-associated herpesvirus expresses an array of viral microRNAs in latently infected cells. Proc Natl Acad Sci U S A 102:5570-5.
6. Calin, G. A., and C. M. Croce. 2006. MicroRNA signatures in human cancers. Nat Rev Cancer 6:857-66.
7. Cullen, B. R. 2004. Transcription and processing of human microRNA precursors. Mol Cell 16:861-5.
8. Dittmer, D., M. Lagunoff, R. Renne, K. Staskus, A. Haase, and D. Ganem. 1998. A cluster of latently expressed genes in Kaposi's sarcoma-associated herpesvirus. J Virol 72:8309-15.
9. Dittmer, D. P. 2003. Transcription profile of Kaposi's sarcoma-associated herpesvirus in primary Kaposi's sarcoma lesions as determined by real-time PCR arrays. Cancer Res 63:2010-5.
10. Dittmer, D. P., C. M. Gonzalez, W. Vahrson, S. M. DeWire, R. Hines-Boykin, and B. Damania. 2005. Whole-genome transcription profiling of rhesus monkey rhadinovirus. J Virol 79:8637-50.
11. Duan, R., C. Pak, and P. Jin. 2007. Single nucleotide polymorphism associated with mature miR-125a alters the processing of pri-miRNA. Hum Mol Genet 16:1124-31.
12. Eisen, M. B., P. T. Spellman, P. O. Brown, and D. Botstein. 1998. Cluster analysis and display of genome-wide expression patterns. Proc Natl Acad Sci U S A 95:14863-8.
13. Gaur, A., D. A. Jewell, Y. Liang, D. Ridzon, J. H. Moore, C. Chen, V. R. Ambros, and M. A. Israel. 2007. Characterization of microRNA expression levels and their biological correlates in human cancer cell lines. Cancer Res 67:2456-68.
14. Gottwein, E., X. Cai, and B. R. Cullen. 2006. A novel assay for viral microRNA function identifies a single nucleotide polymorphism that affects Drosha processing. J Virol 80:5321-6.
15. Griffiths-Jones, S., R. J. Grocock, S. van Dongen, A. Bateman, and A. J. Enright. 2006. miRBase: microRNA sequences, targets and gene nomenclature. Nucleic Acids Res 34:D140-4.
16. Griffiths-Jones, S., H. K. Saini, S. van Dongen, and A. J. Enright. 2008. miRBase: tools for microRNA genomics. Nucleic Acids Res 36:D154-8.
17. Heid, C. A., J. Stevens, K. J. Livak, and P. M. Williams. 1996. Real time quantitative PCR. Genome Res 6:986-94.
18. Hilscher, C., W. Vahrson, and D. P. Dittmer. 2005. Faster quantitative real-time PCR protocols may lose sensitivity and show increased variability. Nucleic Acids Res 33:e182.
19. Jiang, J., E. J. Lee, Y. Gusev, and T. D. Schmittgen. 2005. Real-time expression profiling of microRNA precursors in human cancer cell lines. Nucleic Acids Res 33:5394-403.

20. Kirshner, J. R., K. Staskus, A. Haase, M. Lagunoff, and D. Ganem. 1999. Expression of the open reading frame 74 (G-protein-coupled receptor) gene of Kaposi's sarcoma (KS)-associated herpesvirus: implications for KS pathogenesis. J Virol 73:6006-14.
21. Landgraf, P., M. Rusu, R. Sheridan, A. Sewer, N. Iovino, A. Aravin, S. Pfeffer, A. Rice, A. O. Kamphorst, M. Landthaler, C. Lin, N. D. Socci, L. Hermida, V. Fulci, S. Chiaretti, R. Foa, J. Schliwka, U. Fuchs, A. Novosel, R. U. Muller, B. Schermer, U. Bissels, J. Inman, Q. Phan, M. Chien, D. B. Weir, R. Choksi, G. De Vita, D. Frezzetti, H. I. Trompeter, V. Hornung, G. Teng, G. Hartmann, M. Palkovits, R. Di Lauro, P. Wernet, G. Macino, C. E. Rogler, J. W. Nagle, J. Ju, F. N. Papavasiliou, T. Benzing, P. Lichter, W. Tam, M. J. Brownstein, A. Bosio, A. Borkhardt, J. J. Russo, C. Sander, M. Zavolan, and T. Tuschl. 2007. A mammalian microRNA expression atlas based on small RNA library sequencing. Cell 129:1401-14.
22. Lao, K., N. L. Xu, Y. A. Sun, K. J. Livak, and N. A. Straus. 2006. Real time PCR profiling of 330 human micro-RNAs. Biotechnol J.
23. Lee, E. J., M. Baek, Y. Gusev, D. J. Brackett, G. J. Nuovo, and T. D. Schmittgen. 2008. Systematic evaluation of microRNA processing patterns in tissues, cell lines, and tumors. RNA 14:35-42.
24. Lee, E. J., Y. Gusev, J. Jiang, G. J. Nuovo, M. R. Lerner, W. L. Frankel, D. L. Morgan, R. G. Postier, D. J. Brackett, and T. D. Schmittgen. 2007. Expression profiling identifies microRNA signature in pancreatic cancer. Int J Cancer 120:1046-54.
25. Lomeli, H., Tygai, S., Pritchard, C.G., Lizardi, P.M., and Kramer, F.R. 1989. Quantitative assays based on the use of replicable hybridization probes. Clinical Chemistry 35:1826-1831.
26. Mackay, I. M., K. E. Arden, and A. Nitsche. 2002. Real-time PCR in virology. Nucleic Acids Res 30:1292-305.
27. Morlando, M., M. Ballarino, N. Gromak, F. Pagano, I. Bozzoni, and N. J. Proudfoot. 2008. Primary microRNA transcripts are processed co-transcriptionally. Nat Struct Mol Biol.
28. Mullis, K. B., and F. A. Faloona. 1987. Specific synthesis of DNA in vitro via a polymerase-catalyzed chain reaction. Methods Enzymol 155:335-50.
29. O'Hara, A. J., W. Vahrson, and D. P. Dittmer. 2008. Gene alteration and precursor and mature microRNA transcription changes contribute to the miRNA signature of primary effusion lymphoma. Blood 111:2347-53.
30. Papin, J., W. Vahrson, R. Hines-Boykin, and D. P. Dittmer. 2004. Real-time quantitative PCR analysis of viral transcription. Methods Mol Biol 292:449-80.
31. Pfaffl, M. W. 2001. A new mathematical model for relative quantification in real-time RT-PCR. Nucleic Acids Res 29:E45-5.
32. Pfaffl, M. W., G. W. Horgan, and L. Dempfle. 2002. Relative expression software tool (REST) for group-wise comparison and statistical analysis of relative expression results in real-time PCR. Nucleic Acids Res 30:e36.
33. Pfeffer, S., A. Sewer, M. Lagos-Quintana, R. Sheridan, C. Sander, F. A. Grasser, L. F. van Dyk, C. K. Ho, S. Shuman, M. Chien, J. J. Russo, J. Ju, G. Randall, B. D. Lindenbach, C. M. Rice, V. Simon, D. D. Ho, M. Zavolan, and T. Tuschl. 2005. Identification of microRNAs of the herpesvirus family. Nat Methods 2:269-76.
34. Rice, P., I. Longden, and A. Bleasby. 2000. EMBOSS: the European Molecular Biology Open Software Suite. Trends Genet 16:276-7.
35. Rozen, S. a. S., H.J.. 1998, posting date. Primer3 Software Distribution. [Online].
36. Samols, M. A., J. Hu, R. L. Skalsky, and R. Renne. 2005. Cloning and identification of a microRNA cluster within the latency-associated region of Kaposi's sarcoma-associated herpesvirus. J Virol 79:9301-5.
37. Schmittgen, T. D., J. Jiang, Q. Liu, and L. Yang. 2004. A high-throughput method to monitor the expression of microRNA precursors. Nucleic Acids Res 32:e43.

38. Schmittgen, T. D., E. J. Lee, J. Jiang, A. Sarkar, L. Yang, T. S. Elton, and C. Chen. 2008. Real-time PCR quantification of precursor and mature microRNA. Methods 44:31-8.
39. Simon, R. M., E. L. Korn, L. M. McShane, M. D. Radmacher, G. W. Wright, and Y. Zhao (ed.). 2003. Design and Analysis of DNA Microarray Investigations. Springer, New York.
40. Storey, J. D., and R. Tibshirani. 2003. Statistical significance for genomewide studies. Proc Natl Acad Sci U S A 100:9440-5.
41. Vandesompele, J., K. De Preter, F. Pattyn, B. Poppe, N. Van Roy, A. De Paepe, and F. Speleman. 2002. Accurate normalization of real-time quantitative RT-PCR data by geometric averaging of multiple internal control genes. Genome Biol 3:RESEARCH0034.
42. Volinia, S., G. A. Calin, C. G. Liu, S. Ambs, A. Cimmino, F. Petrocca, R. Visone, M. Iorio, C. Roldo, M. Ferracin, R. L. Prueitt, N. Yanaihara, G. Lanza, A. Scarpa, A. Vecchione, M. Negrini, C. C. Harris, and C. M. Croce. 2006. A microRNA expression signature of human solid tumors defines cancer gene targets. Proc Natl Acad Sci U S A 103:2257-61.
43. Xia, T., A. O'Hara, I. Araujo, J. Barreto, E. Carvalho, J. B. Sapucaia, J. C. Ramos, E. Luz, C. Pedroso, M. Manrique, N. L. Toomey, C. Brites, D. P. Dittmer, and W. J. Harrington, Jr. 2008. EBV microRNAs in primary lymphomas and targeting of CXCL-11 by ebv-mir-BHRF1-3. Cancer Res 68:1436-42.
44. Yanaihara, N., N. Caplen, E. Bowman, M. Seike, K. Kumamoto, M. Yi, R. M. Stephens, A. Okamoto, J. Yokota, T. Tanaka, G. A. Calin, C. G. Liu, C. M. Croce, and C. C. Harris. 2006. Unique microRNA molecular profiles in lung cancer diagnosis and prognosis. Cancer Cell 9:189-98.
45. Yi, R., B. P. Doehle, Y. Qin, I. G. Macara, and B. R. Cullen. 2005. Overexpression of exportin 5 enhances RNA interference mediated by short hairpin RNAs and microRNAs. Rna 11:220-6.
46. Zeng, Y., R. Yi, and B. R. Cullen. 2005. Recognition and cleavage of primary microRNA precursors by the nuclear processing enzyme Drosha. Embo J 24:138-48.

CHAPTER 4

MASSIVELY PARALLEL MICRORNA PROFILING IN THE HAEMATOLOGIC MALIGNANCIES

Ryan D. Morin, Florian Kuchenbauer, R. Keith Humphries and Marco A. Marra

BC Genome Sciences Centre, BC Cancer Agency, Vancouver, Canada
Terry Fox Laboratory, BC Cancer Agency, Vancouver, Canada

MicroRNAs are important regulators with various roles in differentiation and development including the complex process of hematopoeisis. MicroRNAs have been implicated in the development and progression of various cancers including hematologic malignancies. Many of these are also linked to normal hematopoeisis. MicroRNA profiling technologies have evolved over time and the current state of the art (massively parallel microRNA sequencing, or miRNA-seq) enables deep profiling of the entire microRNAome. MiRNA-seq profiles can differentiate patient biopsies from clinically distinct subtypes of diffuse large B-cell lymphoma. The combination of miRNA-seq and transcriptome sequencing (RNA-seq) will offer the potential to identify novel miRNA-target interactions as well as polymorphisms and mutations (in genes or micoRNAs) that alter microRNA targeting during the development of cancer.

1. Introduction

MicroRNAs are recently discovered regulatory RNA molecules that affect the dosage of important genes involved in development, cellular differentiation and apoptosis.[1] Dysregulation of microRNAs can result in devastating effects to the cell and lead to diseases such as cancer.[2,3] For many years, the study of mature microRNAs involved laborious techniques such as cloning (for discovery) and Northern blots (for measuring expression). A major advancement in microRNA expression profiling was the parallelization of microRNA detection by situating complementary probes on a fixed surface such as microarray slides[4] or microscopic beads.[5]

These techniques enabled rapid, routine and relatively inexpensive quantification of microRNA expression and have resulted in a virtual flood of data in the field.[5-8] Despite these advances, these techniques delivered only limited insight into the microRNA transcriptome. Direct sequencing of small RNAs (including microRNAs) offered important advantages over hybridization-based methodologies for measuring microRNA abundance.

With the advent of massively parallel sequencing methodologies, it became generally feasible to measure the expression of the entire microRNAome of a sample.[9] In addition to revealing microRNA abundance with a larger dynamic range than microarrays, sequencing allows the discovery of new microRNA genes that have previously evaded discovery.[9,10] Deep sequencing of microRNA libraries provides the prospect of detecting polymorphisms, somatic mutations and RNA edits in mature microRNAs, all of which have implications for human physiology and pathophysiology.

An ongoing complication in studying human microRNAs is that the recognition of microRNA targets often involves limited or "relaxed" complementarity.[11] In response to this, new and improved target prediction algorithms are continually being released, with numerous emerging approaches that rely on differing methodologies and assumptions. One major advance in this field has been the use of messenger RNA (mRNA) microarrays to identify the transcripts that become depleted when a given microRNA is over-expressed.[8] Studies employing mRNA expression in combination with microRNA expression have helped to verify or refute suspected microRNA/mRNA targeting interactions.[6,8,12] Massively parallel sequencing has also been applied to sequence messenger RNAs using a technique termed RNA-seq.[13-15] Because this new approach reveals the sequence of mRNAs as well as their abundance, RNA-seq provides a richer data source compared to mRNA microarrays. It follows that combined microRNA and mRNA profiling using massively parallel sequencing will yield unprecedented opportunities for bioinformatic analysis that should provide novel insights into the targets and functions of microRNAs. This chapter will discuss current state-of-the-art approaches for microRNA sequencing and profiling. We will highlight the benefits and potential opportunities offered by massively parallel microRNA sequencing (miRNA-seq) and mRNA sequencing (RNA-seq) in further understanding cancer with a focus on the well characterized hematologic malignancies.

2. miRNA-seq: Massively Parallel MicroRNA Sequencing

Until recently, microRNA microarrays were the only cost effective way to simultaneously measure the expression of many microRNAs. Array experiments yield reproducible results that facilitate classification of cancers by virtue of similar microRNA expression profiles[16] while also enabling differentiation of subclasses of diagnostically similar (yet clinically distinct) cancers.[17] Interestingly, it has been shown that an expression profile of 217 microRNAs can better distinguish cancers of different subtypes and stages of differentiation than ~16,000 protein-coding genes from the same samples.[5] Two recent studies have shown a similar result in the ability to differentiate the two common subtypes of diffuse large B-cell lymphoma (DLBCL), albeit using cell line models of the disease.[18]

Though arrays have enabled pivotal discoveries in a many normal biological processes and diseases, their use and interpretation can be challenged by technical limitations. For example, unlike standard mRNA microarrays, there are limited options for probe design. Confounding this issue are the numerous microRNA families encoding highly similar mature sequences. The incorporation of Locked Nucleic Acid (LNA) nucleotide analogs into probes provided a considerable enhancement in the robustness of these technologies, potentially facilitating discrimination between microRNAs with single nucleotide differences.[19]

Despite ongoing improvements to microRNA microarrays, they are inherently restricted to detection and profiling the known microRNA or predicted sequences identified by sequencing or homology searches.[20] Another limitation arises from the observation that mature microRNAs have been shown to exhibit sequence variability owing to polymorphisms, mutations and enzymatic modifications[9,21] It is possible that any of these modifications could affect the efficacy of any hybridization-based approach for measuring microRNA abundance. A rather predominant modification that appears to affect most (if not all) microRNAs is the addition of un-templated 3' nucleotides.[9] In a recent study, we conducted experiments that suggested that stem-loop quantitative RT-PCR primers can measure only a fraction of the intended complementary microRNA in a sample due to these additions (See Figure 1 for an example).[22]

This may not be a problem as long as these additions affect all microRNAs equally. However, if there are sequence alterations that affect certain molecules (for example, a polymorphism or mutation), an assay may erroneously report this as a reduction in the expression level of that microRNA. Indeed, some small RNA libraries we have sequenced contain considerable representation of the non-reference allele of a known SNP (data not shown). These polymorphisms may even be missed if the sequence data is not carefully aligned with appropriate software and sequence errors and alignment artifacts removed where possible.

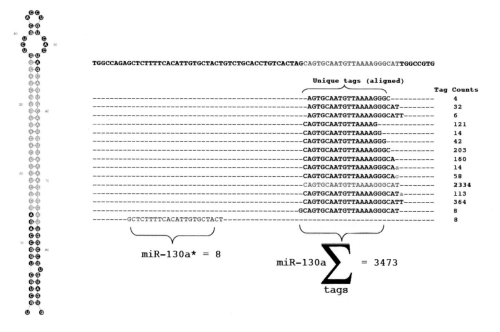

Figure 1. MicroRNA profiling by massively parallel sequencing (miRNA-seq). Short reads deriving from small RNAs are first aligned to the appropriate reference genome (after any preprocessing such as adaptor sequence removal). The aligned reads are counted (right) and the sum of all distinct reads unambiguously derived from a single miRNA reflects the abundance/expression of that microRNA.

Clearly, another benefit of sequencing is that it provides the potential to discover novel microRNA genes. Though many uncharacterized small RNAs are observed in sequencing libraries, a large number of these do not appear to derive from transcripts able to form stable substrates for Drosha and Dicer, key enzymes involved in microRNA maturation. These include a recently uncovered class of microRNA termed miRtrons, whose maturation does not follow that of typical microRNAs.[23] Application of rigorous approaches involving RNA folding and machine learning-based classifiers has resulted in the discovery of many new human microRNA genes in the past few years. These approaches typically rely on small RNA sequence data to identify candidate microRNA genes[9,10,24] Given that only a few tissue types have thus far been profiled using deep sequencing strategies, many more novel microRNAs likely remain to be annotated. Though it is reasonable to assume that most of the highly conserved microRNAs have already been identified by computational means, we speculate that undiscovered non-conserved microRNA genes might play important roles in human disease.

The quantitative detection of microRNA molecules via sequencing only recently became cost effective. Previously, microRNA sequences could only be obtained by capturing small RNAs (using single stranded ligation) followed by reverse transcription, cloning and sequencing of individual cDNAs using capillary sequencing. Later, a procedure that parallels serial analysis of gene expression ("miRAGE") was introduced to make sequencing more cost effective, but the gain in throughput was incremental (i.e. a few dozen, rather than a single microRNA could be sequenced in a single sequencing reaction).[25] Owing to rapid developments in both sequencing chemistry and throughput it is now possible to conduct a thorough interrogation of the microRNAome of individual biological samples without the need to concatamerize small RNAs (a limitation of the miRAGE approach).

Massively parallel sequencing strategies allow the simultaneous 'reading' of sequences of up to millions of cDNAs derived from small RNA fragments. The first such approach (termed 454) is based on pyrosequencing technology.[26] In essence, this method involves the concurrent synthesis of cDNA molecules representing small RNAs on hundreds of thousands of beads. Each of the four nucleoside triphosphates are introduced (in series) and each bead on which that nucleotide is incorporated produces visible light (due to a chemical reaction involving the pyrophospate byproduct). This process is repeated for over 100 cycles. By capturing the signal from each bead during these cycles, a read-out of signal intensity results in a sequence "read" from each bead representing the sequence of the original template molecule. This technology was the first step towards routine profiling of the entire microRNAome of a single sample.

A more recently available technology known as Illumina (or Solexa) sequencing is capable of producing read numbers that are orders of magnitude greater.[27] Rather than using beads, the Illumina technology involves universal primers that are physically anchored to a fixed surface inside a "flow cell" (Figure 2). A sequence-ready library (with flanking sequences complementary to the universal primers) is introduced into a fresh flowcell and bridge amplification produces clusters comprising double-stranded DNA each derived from a single template molecule. As in 454 sequencing, each cluster is individually sequenced in parallel by successive addition of the four nucleoside triphospates. This process is distinct in that the nucleotides have a 3' reversible terminator moiety (which is actually the fluorophore).

This effectively prevents successive incorporation of multiple nucleotides in polynucleotide tracts, a major issue with early 454 sequencing. Images of the laser-excited fluorophores are captured after each nucleotide addition step and an

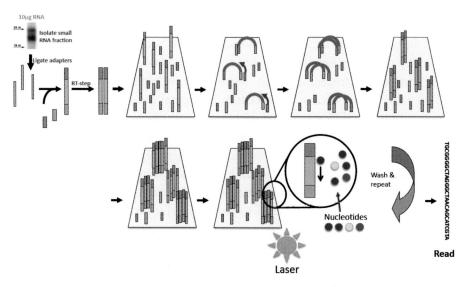

Figure 2. Illumina massively parallel sequencing. After total RNA is size-fractionated to retrieve the desired size range, universal sequencing adaptors are ligated on to either end of the small RNAs. The complementary strand is synthesized (by reverse-transcription) resulting in a sequence-ready cDNA library. This library is applied to a flow-cell and millions of clusters are generated by bridge amplification (red arrow). The primers for bridge-amplification (blue/violet) are complementary to the universal sequencing adaptor. Clusters are sequenced by successive additions of fluorescent ddNTPs. Images from each step are overlaid to extract a sequence read from every cluster on the flowcell.

overlay of all images is used to produce full-length sequence reads. Individual reads produced by this technology are inherently shorter (currently 36-50nt) but this is not a concern in small RNA applications as the read length need only exceed the length of a microRNA. Because the diversity of the microRNAome is much less than the number of reads, many identical sequences are produced (and can be considered multiple observations).

By considering each read as a single observation of a molecule of that microRNA, massively parallel sequence data provides the expression profile of a sample while offering the potential to reveal sequence discrepancies between microRNAs and the human reference genome. However, with only a few exceptions,[9,21] microRNA reads that do not match the genome perfectly are generally discarded due to difficulties in alignment and separating the real observations from noise arising from difficulties in alignment to the human reference genome or sequencing errors. This mindset was encouraged by the knowledge that each individual sequence read had a high error rate (relative to classical capillary sequence data). A set of aligned reads corresponding to a single microRNA gene is shown in Figure 1. Next-generation alignment software such as SOAP or Novoalign are capable of finding optimal alignments in the human genome while allowing multiple mismatches and variable read length.[28] Considering that the variability of microRNAs at the sequence level is poorly understood, routine global analysis of reads (including those with imperfect alignments) is important for proper quantitation of microRNAs and for discovery of known or novel sequence variants.

Profiling microRNAs by sequencing is not subject to the same limitations as hybridization-based methods such as microarrays or real-time PCR. This approach also has a larger dynamic range than array-based technologies and has proven powerful in revealing both large and subtle changes in microRNA abundance between samples.[22] Recent studies have shown that microRNAs circulating in the plasma can be profiled using this approach and it has been shown that serum microRNA levels can serve as prognostic indicators in various cancers.[29,30] Our data shows that miRNA-seq data is equivalent, if not better at classifying subtypes of diffuse large B-cell lymphoma (Figure 3A).

Previously, classification of DLBCL using microRNAs had only been accomplished using cell line models of the disease.[18] This chapter includes data from patient biopsies showing that miRNA-seq can recapitulate these findings. Taken together with the added benefit of revealing novel molecules and nucleotide substitutions caused by mutation, polymorphism or RNA editing, sequence based microRNA expression profiling promises to yield important insights into diseases such as cancer. This suggests the possibility of developing powerful diagnostics that rely on the presence of key microRNAs (or microRNAs with specific mutations) in the blood.

3. RNA-seq: Massively Parallel Messenger RNA Sequencing

Until recently, there were two basic options for measuring the expression profile of a sample: microarrays and serial analysis of gene expression (SAGE). The former was the preferred choice for many as the sequencing cost to produce a "deep" SAGE library was generally restrictive. Shortly after the miRNA-seq technique was published, a comparable approach was developed that allowed profiling entire transcripts using massively parallel sequencing.[31,15,14] Termed RNA-seq, this method marries the digital nature of SAGE with the full-transcript coverage abilities of exon tiling arrays.[15] In simple terms, RNA-seq involves "shotgun" sequencing entire transcripts in parallel using short reads. Because of this, added benefits of RNA-seq include the ability to identify single base changes (i.e. SNPs or mutations) and alternative splicing events, both of which

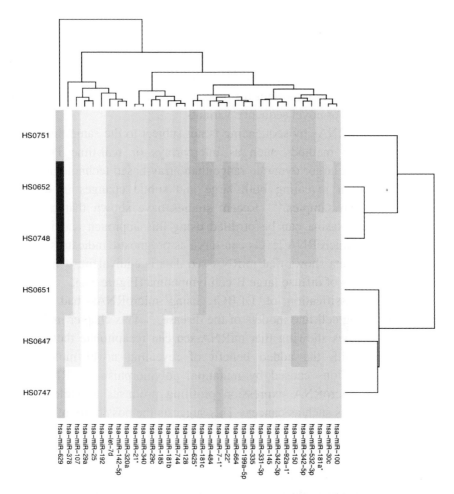

Figure 3A. Measuring abundance of mature miRNAs and pre-miRNAs in DLBCL samples. These heat maps summarize the microRNA profile of a number of DLBCL biopsy samples as measured by miRNA-seq (A) with red indicating low abundance and yellow indicating high abundance. These microRNAs were chosen based on their level of differential expression between libraries based on DLBCL subtype (A). Notably, hierarchical clustering using only these microRNAs (A) correctly classified these samples by DLBCL subtype (See comparable data from RNA-seq using 150 genes). The first three samples were classified as ABC using mRNA expression levels whereas the latter 3 were classified as GCB. Though most microRNAs show the same profile across all patients, a few are distinct in a given sample. There are two different sets of microRNAs that have previously been described as able to separate ABC and GCB subtypes of DLBCL[18,64] and interestingly, neither is able to distinguish the two subtypes using this data. This may reflect a difference in the technology (arrays vs. sequencing) or differences in miRNA expression in cell lines as opposed to primary biopsy samples.

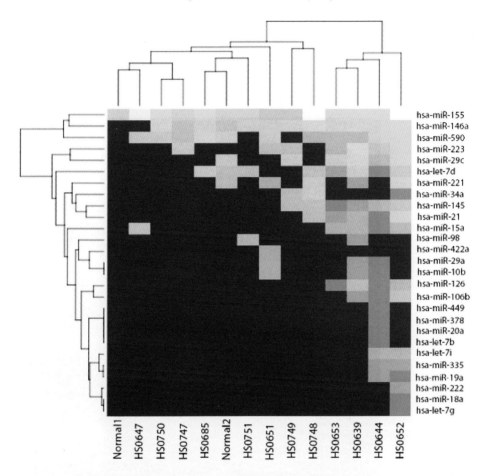

Figure 3B. Measuring abundance of mature miRNAs and pre-miRNAs in DLBCL samples. These heat maps summarize the microRNA profile of a number of DLBCL biopsy samples as measured by mRNA-seq (B) with red indicating low abundance and yellow indicating high abundance. These microRNAs were chosen based on their level of differential expression between libraries based on their known involvement in cancer (B).

can affect the interaction between a microRNA and its target.[32,33] In Figure 4, we show that, much like the miRNA-seq data, the expression values obtained from RNA-seq are able to differentiate DLBCL patient samples. In this case, a much larger number of genes are required to correctly separate DLBCL samples into the correct subgroups.

The RNA-seq approach is complementary to miRNA-seq for many reasons, some of which may not be immediately apparent. By enumerating unambiguously aligned reads, one can quantitate the expression of protein

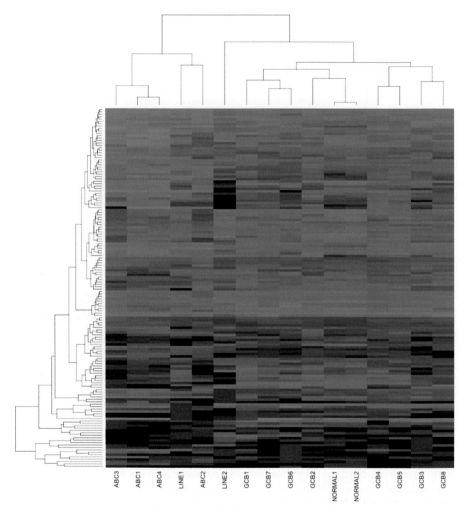

Figure 4. Classification of DLBCL patient samples (and cell lines) into subtypes using RNA-seq gene expression data. The gene expression profile of a set of 185 protein coding genes have been shown to allow differentiation of DLBCL subtypes.[65] This figure demonstrates that using RNA-seq libraries from patient samples recapitulates this. All ABC-subtype (ABC 1-4) and one DLBCL cell line (LINE1) cluster together, while all GCB-subtype samples (GCB 1-8) cluster with germinal centre B-cells (NORMAL 1 and 2).

coding transcripts (in the case of RNA-seq) and mature miRNAs (in the case of miRNA-seq). As with array-based expression measurements, one should be able to identify inverse correlations between microRNAs and their predicted targets, resulting from the degradation of mRNAs targeted by the RISC complex.

The enhanced dynamic range and sensitivity of these approaches over microarrays suggests that more miRNA/target interactions should be detectable from experiments with comparable numbers of samples. Interestingly, we have observed that RNA-seq libraries generally contain reads from many microRNA genes. The fact that the abundance of these reads does not generally correlate with mature microRNA abundance (Figure 3B) and that they derive from the entire miRNA gene loci suggests that they derive from unprocessed pre-microRNAs that may not yield mature microRNAs. This may reflect inefficiencies in the microRNA biogenesis process or the effect of aberrations blocking this process. When a sufficient number of reads is available, these sequence alignments can reveal polymorphisms, mutations and RNA edits (all of which are known to affect pre-microRNA processing). MiR-146a has a well-characterized polymorphism known to affect its maturation. This microRNA is one of the most abundant and consistently observed across RNA-seq libraries. If the RNA-seq reads provide a measure of incomplete maturation of pre-miRNAs, a supposition that is unproven, then the reason for absence of a microRNA in a sample (RNA editing, polymorphism or mutation) may be revealed within these reads.

4. MicroRNAs and Hematopoiesis

MicroRNAs have been identified in diverse species ranging from plants to mammals and exhibit sequence conservation spanning millions of years of evolution. This, along with the observation that certain microRNAs exhibit tissue-specific and/or developmentally regulated expression, point to the role of microRNAs in complex biological processes such as cellular differentiation. This is well illustrated from the already extensive data now accumulated for important roles of microRNAs in both normal and malignant hematopoietic processes. This knowledge, along with the role of key malignancy-associated microRNAs in other cancers are summarized in Figure 5, Table 1 and discussed in the following section.

Mammalian hematopoiesis has been extensively characterized in mice, with pioneering work by Chen et al. demonstrating the presence and relevance of microRNAs in this process.[34] In this work, more than 100 microRNAs were cloned from mouse bone marrow and their expression in different hematopoietic tissues validated. MiR-181, miR-223 and miR-142 were found to have distinct expression patterns in bone marrow lineage cells. Mature miR-181 expression was detected in progenitor cells and B-lymphocytes. In contrast, miR-223 revealed a more myeloid confined expression, whereas miR-142 exhibited a less specific expression pattern in erythroid, lymphoid and myeloid cells.

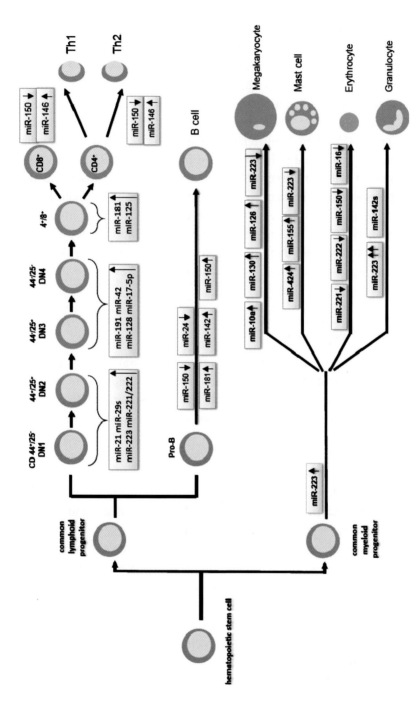

Figure 5. Differential expression of miRNAs during hematopoeisis. This figure summarizes current functional as well as expression studies of miRNAs during hematopoietic differentiation. ↑, upregulation; ↓, downregulation.

Interestingly, supporting their conserved function, microRNA expression profiling in human lineage cells showed comparable results.[35] Functional studies of the effect of ectopic expression of these microRNAs in hematopoietic progenitor cells revealed substantial alterations in lineage differentiation.

For example, increased miR-181 levels resulted in a concomitant increase of B-lymphoid cells, whereas ectopic expression of miR-142 affected the T-lymphoid cells.[34] This initial study affirmed the importance of microRNAs in hematopoiesis and, more specifically, that genes regulated by miR-181 are positive regulators of B-cell differentiation.

As we have previously discussed, concurrent processing of microRNA and messenger RNA along with in-silico target predictions can reveal biologically relevant microRNA/mRNA regulatory relationships. This approach was used by Georgantas et al. to formulate a model of hematopoietic differentiation regulated by microRNAs.[36] In total, the authors found 33 microRNAs expressed in CD34+ hematopoietic stem-progenitor cells from normal human bone marrow and mobilized human peripheral blood stem cell harvests. MiR-181a and miR-128 were predicted to inhibit differentiation of all hematopoietic lineages by regulating molecules critical to very early steps in hematopoiesis. Other microRNAs such as miR-146 were predicted to regulate general lymphoid differentiation at the progenitor level. In contrast, miR-155, miR-24a and miR-17 were shown to be potential regulators of myeloid progenitors. MiR-146, miR-155 and miR-17 have all been associated with various cancers (hematologic and otherwise) and discussion of these microRNAs will be revisited later in this chapter.

Although this study predicted microRNAs to be regulators in early stages of hematopoietic differentiation, it did not address whether microRNAs were equally important for the terminal differentiation of progenitor cells to the multitude of blood cell types. The requirement for microRNAs in T-cell development was first addressed by generating conditional Dicer knockout mouse models[37] in which microRNA generation would be blocked or severely impaired. In these studies conditional Dicer constructs were created to suppress Dicer function during an early and a later stage of T cell development. Knocking-out Dicer in the early stage of thymocyte development impaired the survival of T-cells. Loss of Dicer function in late T-cell development impaired cytotoxic T-cell development and exhibited aberrant T-helper cell differentiation. These results indicated that Dicer is necessary for T-cell differentiation with the corollary that microRNAs are essential in this process. A comprehensive follow-up to this study was performed by Neilson et al., measuring the expression of microRNAs by cloning and sequencing from 6 stages of T-cell development.[38]

This study revealed distinct microRNA expression patterns in thymocyte maturation from various double-negative (DN) stages to mature CD4+ and CD8+ cells and suggested a significant regulatory role for miR-181 also in T-cell development.

A distinct set of microRNAs are important in the latter stages of B-cell maturation, including those deriving from a polycistronic transcript known as the miR-17-92 cluster. Early evidence for the importance of these microRNAs in B-cells stemmed from the observation that this locus is frequently amplified (and highly expressed) in human B-cell lymphomas.[3] Targeted deletion of the miR-17-92 cluster directly demonstrated its role in B-cell development. Knocking out this cluster in mice inhibits the pro-B to pre-B transition and has been associated with increased levels of Bim (a pro-apoptotic protein). These experiments suggest that the miR-17-92 cluster is crucial for the transition from pre-B to pro-B lymphocyte development, enhancing the survival of the B-cells at this stage by targeting Bim. This also points to the oncogenic role of this microRNA cluster because Bim is a regulator of BCL2 (a pro-survival gene commonly over-expressed in lymphomas).

Many of the microRNAs discussed so far are expressed in various tissues and, it follows, are involved in other processes besides hematopoiesis. A recent cloning-based survey of microRNAs from more than 250 tissue libraries surprisingly identified only 5 microRNAs that were highly specific for hematopoietic cells.[39] These were miR-142, miR-144, miR-150, miR-155, and miR-223. In contrast to miR-150, which is mainly expressed in the lymphoid compartment and miR-223, which is typically expressed in mature myeloid cells, miR-142 expression does not follow any specific expression pattern and is expressed in all hematopoietic cells. MiR-144 showed a general weak expression in myeloid and lymphoid cells and miR-155 has its highest expression in primitive CD34+ cells as well as lymphoid cells, suggesting a role in hematopoietic stem cells and the lymphoid system.

It remains speculative why only a handful of microRNAs display tissue-specific expression whereas several hundred microRNAs can be detected across very distinct tissues and cells types from T-cells to embryonic stem cells. Those widely-expressed microRNAs are more likely to affect their targets in a combinatorial fashion, whereby their effect is a function of a multitude of microRNAs with shared targets. Hence, the expression of individual microRNAs is often not as important as the global expression profile of all microRNAs as well as the target genes of these microRNAs. It still remains to be shown through knockout studies what the roles are for many of the tissue-specific microRNAs. Two such studies, using mice lacking miR-155, have reported defective humoral responses after immunization, related to impaired germinal center formation and to result in reduced antibody class switching to immunoglobulin G1. Furthermore, it seems as miR-155 regulates T cell lineage fate by promoting T helper type 1 versus T helper type 2 differentiation, possibly by targeting the transcription factor c-MAF.

In an effort to test the influence of miR-155 on hematopoietic cell development, miR-155 was constitutively retrovirally expressed in murine progenitor and stem cells.[40] Sustained expression of miR-155 had profound effects on hematopoietic populations, resulting in a myeloproliferative disorder, displaying an increase of Gr-1/Mac-1++ cells after at least 8 weeks of engraftment. However, it was not clear if these mice developed AML over time as a possible long-term consequence from enhanced myeloid cell output. In vitro transduction of miR-155 in human CD34+ cells led to decreased myeloid and erythroid colony formation, only partially mimicking the profound effects seen in the murine system.[36]

These observations demonstrate that misregulated miR-155 expression can affect myeloid or lymphoid development depending on the context of its expression, indicating that miR-155 expression directs hematopoietic cell fate "decisions". Furthermore, these experiments demonstrated that overexpression of miR-155 may be an early event in leukemogenesis.[40] This is an important consideration, because miR-155 is highly expressed in a variety of lymphomas and leukemias (Table 1). Considering the general lack of lineage-specific microRNAs, it would not be surprising to learn that miR-155 and other microRNAs implicated in hematologic malignancies have multiple targets that provide distinct mechanisms of pathology.

A somewhat distinct set of microRNAs appears to be important in maturation of erythocytes and megakaryocytes. In vitro differentiation of human CD34+ cells along the megakaryocyte lineage led to the identification of 20 downregulated microRNAs, including miR-10a, miR-126, miR-106, miR-10b, miR-17 and miR-20.[41] In this study by Garzon et al. miR-130a was found to target MAFB, a factor involved in platelet function.

A similar approach was also used to profile microRNAs during erythropoiesis. In the work of Felli et al. downregulation of miR-221 and miR-222 increased levels of the KIT oncogene. MicroRNA mediated modulation of the KIT receptor led expansion of early erythroblasts.[42] Perhaps surprisingly, dysregulation of miR-221/222 resulting in KIT over-expression is involved in various types of cancers but none of the hematologic malignancies (Table 1). Other microRNAs that have been identified in erythrocyte precursors that

Table 1. MicroRNAs directly associated with oncogenesis. This list includes all microRNAs known to be deregulated in cancer. For each cancer type, a "u" signifies upregulation and "d" downregulation of that microRNA. Unless otherwise stated, these relationships are summarized in a recent review.[66]

microRNA(s)	Cancers (up/down)	Type (Oncogenic/ Tumour Suppressive)
let-7/miR-98	Lung[d], colon[d], ovarian[d], NSCLC[d], adenocarcinoma[d]	TS
miR-9	Hypermethylated in breast carcinoma	TS
miR-10b	Breast cancer	Unknown
miR-15	CLL[d]	TS
miR-16	CLL[d]	TS
miR-21	glioblastoma[u], breast carcinoma[u], lymphoma[u]	TS
miR-26a	Burkitt Lymphoma[u]	O
miR-29	CLL[d], NSCLC[d]	TS
miR-34a	Lung[d], colon[d], neuroblastoma[d], CLL[u],	Both
miR-34b/c	NSCLC[d]	TS
miR-100	Pancreatic[u] [69]	O
miR-122a	HCC[d]	TS
miR-126	metastatic breast carcinoma[d]	TS
miR-142	B-cell leukemias[u]	O
miR-143	B-cell leukemias[d] and lymphomas[d]	TS
miR-145	B-cell malignancies[d], colorectal carcinoma[d] [69]	TS
miR-146	Papillary thyroid carcinoma, Breast, Ovarian, Prostate	TS
miR-155	Leukemias[u], lymphomas[u]	O
miR-195	CLL[u]	Unknown
miR-199a	Hepatocellular carcinoma[d] [69]	TS
miR-221/222	glioblastoma[u], papillary thyroid carcinoma[u], prostate[u], pancreatic[u]	O
miR-223	bladder cancer[u], B-cell lymphoma[u], ovarian cancer[u]	
miR-335	metastatic breast carcinoma[d]	TS
miR-372/373	Testicular cancer[d]	TS
miR-378	B-cell lymphoma[u]	O
miR-106-363 cluster	T-cell leukemias[u], CLL[d]	Both
miR-17-92 cluster	breast[d], lung[u], B-cell lymphoma[u]	O

underwent a progressive downregulation during erythropoiesis were miR-150 and miR-155 combined with upregulation of miR-16 and miR-451 at late stages of differentiation and a biphasic regulation of miR-339 and miR-378.[43] Some of these microRNAs have been implicated in various hematologic malignancies including CLL (miR-16),[44] several cancer cell lines (miR-378)[45] or various leukemias/lymphomas (miR-155).[46]

5. MicroRNAs and Cancer

As discussed above, much evidence now implicates microRNAs as contributing (if not causative) factors in many types of cancers. A provocative observation in this regard was that a large number of the known recurrent genomic alterations involved in cancer are in close proximity to microRNA genes[47] and thus suggesting that these rearrangements affect the expression of microRNAs with tumor suppressive or oncogenic properties. Indeed, there are now numerous reports of microRNA expression differences that are signatures of various neoplasias (compiled in Table 1).

However, large-scale genomic alterations are only one of many mechanisms by which microRNA expression or efficacy can be altered. Likewise, microRNA expression is far from the only mechanism by which oncogenes or tumor suppressors can be deregulated. Numerous well-characterized translocations result in activation of genes involved in apoptosis or cell cycle progression. Some of these oncogenes directly mediate the expression of microRNAs. For example, overexpression of c-Myc leads directly to transcription of the miR-17-92 microRNA cluster (which, in turn, is a posttranscriptional regulator of c-Myc). Hence, microRNA profiling can reveal expression changes that are either directly causative or the result of upstream changes in the development of cancer.

Though large-scale genomic alterations were the first well-characterized mechanisms for microRNA deregulation in cancer, a variety of other mechanisms can produce the same result. Other documented mechanisms altering microRNA expression in cancer include epigenetic inactivation,[48-50] A->I editing of the pre-microRNA,[51] polymorphisms[52,53] or somatic mutations[44,52] which can affect either microRNA transcription or maturation.

Any methodology that measures the abundance of mature microRNAs (i.e. the microRNA expression profile) of cancers can capture differences in abundance, regardless of the underlying cause. Nonetheless, expression profiles may still miss subtle changes that may affect the function of a microRNA. These include the aforementioned sequence-level discrepancies including polymorphisms, somatic mutations and RNA edits,[21,54] all of which have been demonstrated to alter the interaction between a microRNA and its target. An ideal profiling approach would capture differential microRNA expression as well as the mechanism by which expression is altered. Though such a holistic approach does not yet exist, sequencing of small RNA libraries (miRNA-seq) in conjunction with messenger RNA shotgun sequencing (RNA-seq) is likely the most powerful solution offered by today's rapidly evolving technology.

Identifying the microRNAs that are dysregulated in cancers is a first step in a difficult journey. The next major step is to delineate their pathogenic targets. Though many approaches for target prediction exist, the targets of most microRNAs are not known. To date, a total of 40 microRNAs have been experimentally shown to target known cancer genes (Table 2), mainly based on computational predictions validated by the gold standard luciferase assay. There are many more that have been computationally predicted to regulate cancer genes, but determining which of these are real versus false positive targets is laborious.

As we have discussed, the incorporation of mRNA expression data with microRNA expression profiles should enable more accurate identification of true targets of cancer-related microRNAs. One reason for this is that distinct tissues express only a subset of all genes. Hence, even perfect computational target prediction could not determine whether a microRNA and mRNA would be present in the same tissue. This can only be resolved by identifying the pairs of miRNAs and putative targets that are co-expressed across multiple libraries. In addition to this, any microRNA/mRNA interaction that leads to enhanced mRNA degradation will be reflected in a reduction of mRNA in samples enriched for that microRNA. This relationship should manifest as a negative correlation between the abundance of a microRNA and its in-vivo mRNA targets.[55-57]

MicroRNAs can have the functional effect of a tumor suppressor gene or an oncogene. Mir-15 and miR-16 are two microRNAs that act as classical tumor suppressors by down-regulating the anti-apoptotic BCL2 gene in normal B-cells.[58] BCL2 over-expression is observed in multiple B-Cell malignancies including B-CLL and non-Hodgkin's lymphomas. Reduction of miR-15b and/or miR-16 is one of a number of mechanisms leading to BCL2 over-expression and is thought to be the primary mechanism in development of B-CLL.[58] Approximately 20% of follicular lymphomas (FLs) and the majority of de-novo DLBCLs do not harbor an immunoglobulin/BCL2 translocation, a well-known mechanism of BCL2 deregulation in lymphomas. This leads to the appealing hypothesis that miR-15 and miR-16 or other BCL2-targeting microRNAs may have a role in a subset of the B-cell lymphomas.

Table 2. Experimentally validated regulatory interactions between microRNAs and cancer genes. All experimentally validated microRNA/target interactions involving cancer-associated genes are shown. These interactions can be found in the TarBase[67] database and include all genes in the Cancer Gene Census.[68]

microRNA(s)	Target(s)
let-7a, let-7b, let-7g	KRAS, HMGA2, NF2, CCND1, HMGA1, MLLT1
miR-1	ZNF264, LASP1, EGFR, MET, PICALM, TPM3, TPM4
miR-101	MYCN
miR-106a	RB1
miR-124	TRIP11, ELF4, MYH9, CDK6, CDK4
miR-125a	ERBB2
miR-127	BCL6
miR-130	MAFB
miR-133	ERG
miR-15	BCL2, CCND1, WNT3A
miR-155	CBFB, CTNNB1, MET, METTL7A, MSI2, PICALM, TNFRSF10A*
miR-16	BCL2, EGFR, MLLT1, PAFAH1B2, TPM3
miR-17-92	BCL2L11(Bim)*, CDKN1A (p21)*
miR-181	TCL1A
miR-199a*	MET
miR-19, miR-21, miR-214	PTEN
miR-221, miR-222	KIT
miR-23b, miR-24, miR-27b, miR-34	NOTCH1
miR-26a	PLAG1, EZH2
miR-29c	COL1A1
miR-30	CBFB, MET, MLLT1, PAFAH1B2, ZNF294
miR-373	ZNF226, NIN
miR-375	JAK2
miR-378	SUFU, Fus-1
miR-98	HMGA2
miR-150	c-MYB
miR-223	LMO2
miR-146a/b	ROCK1, TRAF6

Another illustrative example of tumor suppressive microRNAs is the repression of RAS oncogenes by let-7, a relationship that is conserved between nematodes and humans.[59] In normal lung tissues, the let-7 microRNAs tightly regulate the expression of KRAS, and the reduced expression of let-7 is a signature of non-small cell (NSC) lung cancers[59] (Table 1). Affirming the importance of this relationship, a common polymorphism in the let-7 binding site of KRAS leads to an increased risk of developing NSC lung cancer.[32] In other words, the efficacy of a tumor-suppressing microRNA can be reduced by misregulation of the microRNA or a reduction of its binding affinity for its target(s). This finding adds a complication to microRNA target prediction, since a somatic mutation in the binding site of a microRNA would have the same effect (on that particular transcript) as a heterozygous knockout of that microRNA. With RNA-seq, now exists the potential to collectively identify SNPs and mutations in candidate miRNA target sites. Moving forward, analysis of paired RNA-seq/miRNA-seq data must concurrently consider the possible binding sites of miRNAs as well as both microRNA and mRNA expression.

Besides the tumor-suppressive microRNAs, there exist a set of microRNAs with oncogenic properties, which are collectively termed "oncomiRs". These include the regulators of the PTEN tumor suppressor such as miR-21. MiR-21 is commonly over-expressed in many cancers including hepatocellular carcinoma (Table 1). In hepatocellular carcinoma cells it has been shown that experimental inhibition of miR-21 can restore PTEN protein and results in decreased cell proliferation.[60]

As we have discussed, another well-studied set of oncogenic microRNAs are those deriving from the miR-17-92 cluster. These cotranscribed microRNAs, which share many common targets, are abundant in hepatocellular carcinoma[61] as well as various classes of B-cell lymphoma.[3,62] Though it has not been directly demonstrated, it is widely accepted that the BIM tumor suppressor gene (BCL2L11) is an important target of these microRNAs. BIM is a transcriptional regulator of BCL2, hence, expression of miR-17-92 indirectly counters the effect of BIM by increasing BCL2 mRNA levels. In line with other oncogenic microRNAs, the over-expression of miR-17-92 results in uncontrolled cellular growth. This appears to be the more prevalent involvement of microRNAs in B-cell lymphomas as opposed to the aforementioned loss of miR-15/miR-16 observed in B-CLL.

Interestingly, not all the B-cell lymphomas with miR-17-92 over-expression have a concomitant drop in BIM (nor do they necessarily exhibit BCL2 over-expression). Recent evidence has implicated CDKN1A (p21) as another target of this microRNA cluster.[63] CDKN1A/p21 is positively regulated (at the transcriptional level) by p53, hence, over-expression of the miR-17-92 cluster can reduce cytoplasmic p21 regardless of p53 status. This finding demonstrates that miR-17-92 over-expression, depending on the expression status of other genes, can lead to lymphoma via very different mechanisms. This further solidifies the importance of concurrently profiling microRNAs and mRNAs in diseased tissue or cell line models. This also highlights the fact that our current understanding of microRNAs in the hematologic malignancies is still evolving and that many other microRNA/gene relationships involved in these diseases have yet to be uncovered.

6. Conclusions

The study of microRNAs has been rapidly gaining momentum with advances in profiling approaches. State of the art technologies now allow us to visualize the expression of microRNAs and messenger RNAs at single base resolution using massively parallel sequencing. For the first time, gene/microRNA expression data can be integrated with knowledge of sequence variations, which may include polymorphisms as well as RNA edits or somatic mutations within microRNAs or their corresponding target sites. This chapter has included some preliminary analysis of patient samples using both RNA-seq and miRNA-seq. We have shown that, as was already known to be the case for microarray-based measurements, mRNA and miRNA expression levels from sequencing data enable us to distinguish subtypes of cancer. Concomitant analysis of gene and microRNA expression data may enable the discovery of novel target interactions, including genes known to be involved in lymphomagenesis. The added information of pre-microRNAs, mature microRNAs and mRNAs at the single base level may reveal somatic mutations that perturb the normal interplay between microRNAs and their targets.

Acknowledgements

This work was supported, in part, by the BC Cancer Foundation and the Terry Fox foundation. This work was supported, in part, by the National Cancer Institute (USA). Florian Kuchenbauer is supported by the Deutsche Forschungsgemeinschaft Germany (grant no. Ku 2288/1-1).

References:

1. Lee, C.T., T. Risom, and W.M. Strauss, MicroRNAs in mammalian development. Birth Defects Res C Embryo Today, 2006. 78(2): p. 129-39.
2. O'Rourke, J.R., M.S. Swanson, and B.D. Harfe, MicroRNAs in mammalian development and tumorigenesis. Birth Defects Res C Embryo Today, 2006. 78(2): p. 172-9.
3. Ota, A., et al., Identification and characterization of a novel gene, C13orf25, as a target for 13q31-q32 amplification in malignant lymphoma. Cancer Res, 2004. 64(9): p. 3087-95.
4. Babak, T., et al., Probing microRNAs with microarrays: tissue specificity and functional inference. Rna, 2004. 10(11): p. 1813-9.
5. Lu, J., et al., MicroRNA expression profiles classify human cancers. Nature, 2005. 435(7043): p. 834-8.
6. Chen, C., et al., Defining embryonic stem cell identity using differentiation-related microRNAs and their potential targets. Mamm Genome, 2007.
7. Zhao, J.J., et al., Genome-wide microRNA profiling in human fetal nervous tissues by oligonucleotide microarray. Childs Nerv Syst, 2006. 22(11): p. 1419-25.
8. Lim, L.P., et al., Microarray analysis shows that some microRNAs downregulate large numbers of target mRNAs. Nature, 2005. 433(7027): p. 769-73.
9. Morin, R.D., et al., Application of massively parallel sequencing to microRNA profiling and discovery in human embryonic stem cells. Genome Res, 2008. 18(4): p. 610-21.
10. Friedlander, M.R., et al., Discovering microRNAs from deep sequencing data using miRDeep. Nat Biotechnol, 2008. 26(4): p. 407-15.
11. Lewis, B.P., C.B. Burge, and D.P. Bartel, Conserved seed pairing, often flanked by adenosines, indicates that thousands of human genes are microRNA targets. Cell, 2005. 120(1): p. 15-20.
12. Giraldez, A.J., et al., Zebrafish MiR-430 promotes deadenylation and clearance of maternal mRNAs. Science, 2006. 312(5770): p. 75-9.
13. Bainbridge, M.N., et al., Analysis of the prostate cancer cell line LNCaP transcriptome using a sequencing-by-synthesis approach. BMC Genomics, 2006. 7: p. 246.
14. Mortazavi, A., et al., Mapping and quantifying mammalian transcriptomes by RNA-Seq. Nat Methods, 2008. 5(7): p. 621-8.
15. Morin, R., et al., Profiling the HeLa S3 transcriptome using randomly primed cDNA and massively parallel short-read sequencing. Biotechniques, 2008. 45(1): p. 81-94.
16. Davison, T.S., C.D. Johnson, and B.F. Andruss, Analyzing micro-RNA expression using microarrays. Methods Enzymol, 2006. 411: p. 14-34.
17. Alizadeh, A.A., et al., Distinct types of diffuse large B-cell lymphoma identified by gene expression profiling. Nature, 2000. 403(6769): p. 503-11.
18. Lawrie, C.H., et al., Microrna expression distinguishes between germinal center B cell-like and activated B cell-like subtypes of diffuse large B cell lymphoma. Int J Cancer, 2007. 121(5): p. 1156-61.
19. Vester, B. and J. Wengel, LNA (locked nucleic acid): high-affinity targeting of complementary RNA and DNA. Biochemistry, 2004. 43(42): p. 13233-41.
20. Berezikov, E., et al., Many novel mammalian microRNA candidates identified by extensive cloning and RAKE analysis. Genome Res, 2006. 16(10): p. 1289-98.
21. Reid, J.G., et al., Mouse let-7 miRNA populations exhibit RNA editing that is constrained in the 5'-seed/ cleavage/anchor regions and stabilize predicted mmu-let-7a:mRNA duplexes. Genome Res, 2008. 18(10): p. 1571-81.
22. Kuchenbauer, F., et al., In depth characterization of the microRNA transcriptome in a leukemia progression model. . Genome Research, 2008. under Review.

23. Ruby, J.G., C.H. Jan, and D.P. Bartel, Intronic microRNA precursors that bypass Drosha processing. Nature, 2007. 448(7149): p. 83-6.
24. Berezikov, E., et al., Diversity of microRNAs in human and chimpanzee brain. Nat Genet, 2006. 38(12): p. 1375-7.
25. Cummins, J.M., et al., The colorectal microRNAome. Proc Natl Acad Sci U S A, 2006. 103(10): p. 3687-92.
26. Margulies, M., et al., Genome sequencing in microfabricated high-density picolitre reactors. Nature, 2005. 437(7057): p. 376-80.
27. Bentley, D.R., et al., Accurate whole human genome sequencing using reversible terminator chemistry. Nature, 2008. 456(7218): p. 53-9.
28. Li, R., et al., SOAP: short oligonucleotide alignment program. Bioinformatics, 2008. 24(5): p. 713-4.
29. Lawrie, C.H., et al., Detection of elevated levels of tumor-associated microRNAs in serum of patients with diffuse large B-cell lymphoma. Br J Haematol, 2008. 141(5): p. 672-5.
30. Chen, X., et al., Characterization of microRNAs in serum: a novel class of biomarkers for diagnosis of cancer and other diseases. Cell Res, 2008. 18(10): p. 997-1006.
31. Marioni, J.C., et al., RNA-seq: an assessment of technical reproducibility and comparison with gene expression arrays. Genome Res, 2008. 18(9): p. 1509-17.
32. Chin, L.J., et al., A SNP in a let-7 microRNA complementary site in the KRAS 3' untranslated region increases non-small cell lung cancer risk. Cancer Res, 2008. 68(20): p. 8535-40.
33. Sandberg, R., et al., Proliferating cells express mRNAs with shortened 3' untranslated regions and fewer microRNA target sites. Science, 2008. 320(5883): p. 1643-7.
34. Chen, C.Z., et al., MicroRNAs modulate hematopoietic lineage differentiation. Science, 2004. 303(5654): p. 83-6.
35. Monticelli, S., et al., MicroRNA profiling of the murine hematopoietic system. Genome Biol, 2005. 6(8): p. R71.
36. Georgantas, R.W., 3rd, et al., CD34+ hematopoietic stem-progenitor cell microRNA expression and function: a circuit diagram of differentiation control. Proc Natl Acad Sci U S A, 2007. 104(8): p. 2750-5.
37. Cobb, B.S., et al., T cell lineage choice and differentiation in the absence of the RNase III enzyme Dicer. J Exp Med, 2005. 201(9): p. 1367-73.
38. Neilson, J.R., et al., Dynamic regulation of miRNA expression in ordered stages of cellular development. Genes Dev, 2007. 21(5): p. 578-89.
39. Landgraf, P., et al., A mammalian microRNA expression atlas based on small RNA library sequencing. Cell, 2007. 129(7): p. 1401-14.
40. Costinean, S., et al., Pre-B cell proliferation and lymphoblastic leukemia/high-grade lymphoma in E(mu)-miR155 transgenic mice. Proc Natl Acad Sci U S A, 2006. 103(18): p. 7024-9.
41. Garzon, R., et al., MicroRNA fingerprints during human megakaryocytopoiesis. Proc Natl Acad Sci U S A, 2006. 103(13): p. 5078-83.
42. Felli, N., et al., MicroRNAs 221 and 222 inhibit normal erythropoiesis and erythroleukemic cell growth via kit receptor down-modulation. Proc Natl Acad Sci U S A, 2005. 102(50): p. 18081-6.
43. Bruchova, H., et al., Regulated expression of microRNAs in normal and polycythemia vera erythropoiesis. Exp Hematol, 2007. 35(11): p. 1657-67.
44. Calin, G.A., et al., A MicroRNA signature associated with prognosis and progression in chronic lymphocytic leukemia. N Engl J Med, 2005. 353(17): p. 1793-801.
45. Jiang, J., et al., Real-time expression profiling of microRNA precursors in human cancer cell lines. Nucleic Acids Res, 2005. 33(17): p. 5394-403.

46. Eis, P.S., et al., Accumulation of miR-155 and BIC RNA in human B cell lymphomas. Proc Natl Acad Sci U S A, 2005. 102(10): p. 3627-32.
47. Calin, G.A., et al., Human microRNA genes are frequently located at fragile sites and genomic regions involved in cancers. Proc Natl Acad Sci U S A, 2004. 101(9): p. 2999-3004.
48. Lehmann, U., et al., Epigenetic inactivation of microRNA gene hsa-mir-9-1 in human breast cancer. J Pathol, 2008. 214(1): p. 17-24.
49. Yang, W., et al., Modulation of microRNA processing and expression through RNA editing by ADAR deaminases. Nat Struct Mol Biol, 2006. 13(1): p. 13-21.
50. Bueno, M.J., et al., Genetic and epigenetic silencing of microRNA-203 enhances ABL1 and BCR-ABL1 oncogene expression. Cancer Cell, 2008. 13(6): p. 496-506.
51. Schmittgen, T.D., REGULATION OF microRNA PROCESSING IN DEVELOPMENT, DIFFERENTIATION AND CANCER. J Cell Mol Med, 2008.
52. Jazdzewski, K., et al., Common SNP in pre-miR-146a decreases mature miR expression and predisposes to papillary thyroid carcinoma. Proc Natl Acad Sci U S A, 2008. 105(20): p. 7269-74.
53. Shen, J., et al., A functional polymorphism in the miR-146a gene and age of familial breast/ovarian cancer diagnosis. Carcinogenesis, 2008. 29(10): p. 1963-6.
54. Kawahara, Y., et al., Redirection of silencing targets by adenosine-to-inosine editing of miRNAs. Science, 2007. 315(5815): p. 1137-40.
55. Farh, K.K., et al., The widespread impact of mammalian MicroRNAs on mRNA repression and evolution. Science, 2005. 310(5755): p. 1817-21.
56. Ruike, Y., et al., Global correlation analysis for micro-RNA and mRNA expression profiles in human cell lines. J Hum Genet, 2008. 53(6): p. 515-23.
57. Nam, S., et al., miRGator: an integrated system for functional annotation of microRNAs. Nucleic Acids Res, 2008. 36(Database issue): p. D159-64.
58. Cimmino, A., et al., miR-15 and miR-16 induce apoptosis by targeting BCL2. Proc Natl Acad Sci U S A, 2005. 102(39): p. 13944-9.
59. Johnson, S.M., et al., RAS is regulated by the let-7 microRNA family. Cell, 2005. 120(5): p. 635-47.
60. Meng, F., et al., MicroRNA-21 regulates expression of the PTEN tumor suppressor gene in human hepatocellular cancer. Gastroenterology, 2007. 133(2): p. 647-58.
61. Connolly, E., et al., Elevated expression of the miR-17-92 polycistron and miR-21 in hepadnavirus-associated hepatocellular carcinoma contributes to the malignant phenotype. Am J Pathol, 2008. 173(3): p. 856-64.
62. Tagawa, H. and M. Seto, A microRNA cluster as a target of genomic amplification in malignant lymphoma. Leukemia, 2005. 19(11): p. 2013-6.
63. Inomata, M., et al., MicroRNA-17-92 downregulates expression of distinct targets in different B-cell lymphoma subtypes. Blood, 2008.
64. Malumbres, R., et al., Differentiation-stage-specific expression of microRNAs in B-lymphocytes and diffuse large B-cell lymphomas. Blood, 2008.
65. Xie, Y. and J.D. Minna, Predicting the future for people with lung cancer. Nat Med, 2008. 14(8): p. 812-3.
66. Schickel, R., et al., MicroRNAs: key players in the immune system, differentiation, tumorigenesis and cell death. Oncogene, 2008. 27(45): p. 5959-74.
67. Sethupathy, P., B. Corda, and A.G. Hatzigeorgiou, TarBase: A comprehensive database of experimentally supported animal microRNA targets. Rna, 2006. 12(2): p. 192-7.
68. Futreal, P.A., et al., A census of human cancer genes. Nat Rev Cancer, 2004. 4(3): p. 177-83.
69. Schmittgen, T.D., Regulation of microRNA processing in development, differentiation and cancer. J Cell Mol Med, 2008. 12(5B): p. 1811-9.

CHAPTER 5

WHO DUNIT? MICRORNAS INVOLVED IN PROSTATE CANCER

Chang Hee Kim

SAIC-Frederick, Inc., NCI-Frederick,
Frederick, MD 21702 Laboratory of Molecular Technology,
Advanced Technology Program

Either as causative agents or responsive elements, microRNAs are differentially expressed in prostate cancer. However, microRNA profiling in prostate cancer has been fraught with discrepancies across studies owing to differences in the profiling technologies used, and in the prostate cancer samples that have been profiled. There is very little overlap although a few microRNAs such as miR-125, miR-100, let-7 seem to recur as differentially expressed microRNAs in prostate tumors compared to normal prostate tissue. The validated targets of some of the differentially expressed microRNAs are either proliferation genes or pro-apoptotic/growth inhibitory genes, suggesting their implication in cancer. Clearly, microRNA profiling technologies need to be better standardized and validated accurately in order to facilitate the discovery of diagnostic and prognostic molecular biomarkers for prostate cancer. microRNAs are also differentially expressed between androgen responsive and androgen refractory prostate cancers, although again there are only a few overlaps across studies. Androgen also seems to regulate the expression of microRNAs. The discovery of a microRNA-mediated androgen signaling pathway may prove invaluable to the treatment of androgen-refractory prostate cancer.

1. Introduction

Prostate cancer (CaP) is the most commonly diagnosed non-cutaneous cancer and the second leading cause of cancer deaths in United States men.[1] 1 in 6 men eventually develop CaP in their lifetime. Microscopic lesions of CaP have been observed in 50% of men 70 to 80 years old.[2] Approximately 186,320 men in the U.S. were diagnosed with CaP in 2008 (www.prostatecancerfoundation.org).[1] The estimated death rate in that year was 28,660.[1] In Europe, the death rate is 80,000 per year.[3] CaP is the second leading cause of cancer death (0.226% of men over 65 years of age) but lung cancer caused by smoking (80%) is 350 times as deadly. As such prostate cancer has been labeled the "indolent cancer" or a "toothless lion". However, CaP still accounts for 3.5% of all male deaths. To put this in perspective, there is a 1/28 lifetime chance of dying from prostate cancer compared to a 1/4000 chance in an automobile accident and a chance 1/100,000 in a plane crash.[4]

Despite intense research efforts, there is much confusion about the different aspects of the disease. Diagnosis and staging of the disease is controversial. Although PSA (Prostate Specific Antigen) and DRE (Digital Rectal Exam) almost always detect the cancer before it becomes metastatic, the benefits of screening large populations are unproven. The PSA test is non-specific[5] and often leads to overdiagnosis (almost as high as 75%)[6] and overly aggressive treatment which may be life-altering.

Clearly, there is a need for better biomarkers to initially diagnose CaP specifically. Even when detected by multiple systematic needle biopsies, it is difficult to determine the size, location and the extent of the lesion. The Gleason grading system is used to distinguish the 5 different stages that prostate cancer cells go through as they change their appearance from normal cells into cancer cells.[7] Upon microscopic examination, the pathologist assigns a Gleason grade to the most common cell pattern and designates a second Gleason grade to the next most common pattern. The two scores are added to arrive at a Gleason score.

Most clinicians believe that biopsies underestimate the stage of the disease and they lack confidence in imaging by magnetic resonance or spectroscopy. Thus, although few prostate cancers are metastatic at diagnosis, physicians are likely to err on the side of caution and include bone scans and computed tomography.

Advanced prostate cancer often responds well to androgen-deprivation therapy, but after about a year or year and a half, many of these cancers become androgen-refractory and become unresponsive to anti-androgen treatment.[8] One possibility is that clones which grow independently emerge upon androgen withdrawal.

The mechanisms by which androgen dependent (AD) prostate cancer cells are converted to androgen independent (AI) prostate (aka hormone refractory prostate cancer (HRPC)) cancer cells must be elucidated in order to find targets for effective drug therapy. In addition, it is critical that biomarkers to distinguish AI prostate cancer from AD prostate cancer be discovered in order to facilitate treatment and to monitor drug response.

Thus, the biological mechanisms responsible for CaP occurrence and progression are largely unknown, and genomic technologies (including mRNA expression profiling and proteomic profiling) have been brought to the frontline of the battle to discover genes and genetic pathways involved in CaP. Recently, microRNAs, which are short (~22 nucleotides or nts) non-coding regulatory RNAs that act post-transcriptionally, have been demonstrated to be involved in cell growth, differentiation and cancer. Bioinformatic analyses predict that the microRNAs may regulate as many as 30% of the human genes.[9]

MicroRNAs are transcribed by Pol II RNA Polymerase into long primary miRNAs (pri-miRNA) presumably when triggered by hormones or other extracellular signals.[10] The several hundred to thousand nucleotide-long pri-miRNA is subsequently processed in the nucleus by the Microprocessor complex (RNAse III enzyme Drosha and a ds DNA binding protein DGCR8/Pasha) into a 60-80 nt stem-loop RNA called the precursor miRNA or pre-miRNA.[11,12,13]

The pre-miRNAs are transported by exportin- 5 in a RanGTP-dependent manner into the cytoplasm[14,15] where a second RNAse III enzyme Dicer cleaves 22 nts from the pre-miRNA to yield a double-stranded miRNA/miRNA*[16] that is finally processed by a helicase[17] activity into the single-stranded mature or active miRNA. One strand of the duplex (the mature or active miRNA) is preferentially incorporated into the RNA-induced Silencing Complex (RISC) where it can act as a guide for RISC to regulate gene expression either by 1) binding with perfect or nearly perfect complementarity to mRNA and causing mRNA degradation or by 2) binding imperfectly with the 3'-UTR sequences to inhibit translation (reviewed in[18]).

Approximately 50% of miRNAs are located at fragile sites on regions amplified or deleted in human cancer.[19] Profiling of microRNAs in various cancers has shown that many miRNAs are aberrantly expressed in cancer cells compared to normal cells, and that microRNA expression profile signatures can be used to distinguish one type or stage of cancer from another accurately. This chapter will review the profiling of microRNA expression in prostate cancer.

In this chapter, I will review the reason why microRNAs are suspect in prostate cancer, the microRNA profiling in CaP (both CaP vs normal and AD CaP vs AI CaP) and the discrepancies that are rampant in this field of research, what we can do to improve the profiling of microRNAs in CaP and finally a hypothesis for the roles that microRNAs may play in CaP based on the functions and the validated targets of the microRNAs that are involved in CaP.

2. Why microRNAs in Prostate Cancer?

The functions of the first two microRNAs that were discovered, lin-4 and let-7 gave clues as to the possible role of microRNAs in cancer. When lin-4 or let-7 is inactivated, specific cells in C. elegans undergo additional cell divisions (proliferation) rather than normal differentiation.[20] Since abnormal cell proliferation is a hallmark of cancers, it seemed plausible that miRNA expression could yield information regarding cancer diagnosis. One obvious reason we believe that microRNAs are involved in CaP is that the few studies of microRNA expression profiling in CaP demonstrate that microRNAs are aberrantly expressed between CaP and normal cells, and between AD and AI prostate cancer cells. Before we delve into the details of microRNA expression profiles, is there any evidence at the DNA level, that microRNAs are involved in prostate cancer?

3. DNA Evidence

There is no one major gene that is mutated in prostate cancer as far as the inherited susceptibility and genetic etiology of the disease is concerned. This is because CaP is 1) very heterogeneous 2) too common: ¼ of men over 63 have lesions in the prostate.[5]

Are there mutations of genes encoding miRNAs in patient tumors? The most compelling evidence that a gene is involved in cancer is the presence of acquired mutations in tumors. The first such example of microRNA genes were miR-15a and miR-16-1 which had chromosomal abnormalities in B cell chronic lymphocytic leukemias (B-CLL).[21] Prior to the discovery of these two microRNAs, the causative genes within the region at chromosome 13q14 deleted in a large fraction of B-CLL could not be found. In fact, in silico studies have shown that 50% of miRNAs are located at fragile sites or regions amplified or deleted in human cancer[19] Comparative Genomic Hybridization (CGH) which detects both gains and losses of DNA copy numbers, has been informative in studying chromosomal aberrations in CaP. In early stage CaP, losses are more common than amplifications or gains, suggesting that tumor suppressors are involved. In contrast, in late-stage metastatic and AI or hormone refractory CaP, gains and amplifications are seen, indicating that oncogenic activation is in play.[5] A study of the association between miRNA locus copy number data obtained using CGH and miRNA expression data in five prostate cancer cell lines and seven xenografts showed significant trend (P = 0.0029) (Porkka 2007). Thus, chromosomal aberrations could cause alterations in miRNA expression.

Recently, a high throuhput CGH array study (Zhang Giannakakis 2006) showed high frequency of genomic aberrations in miRNA loci in human cancers.

Gain of the long arm of chromosome 8 (8q) is one of the most common chromosomal aberrations advanced stage CaP including metastatic and hormone refractory or AI prostate cancer. According to CGH analyses, two minimal regions 8q21 and 8q23-24 have been identified.[5] More recently, Genome Wide Association Studies (GWAS) have shown that multiple independent prostate cancer risk variants are associated with chromosome 8q24.[22] The most commonly deleted chromosomal regions in CaP and 8p and 13q. It remains to be seen whether there are microRNAs in these regions of the genome that are involved in CaP.

Analysis of GWAS SNP variations revealed systematic primary sequence homology/complementarity-driven pattern of associations between disease-linked SNPs, microRNAs and protein coding mRNAs in fifteen different human disorders including prostate cancer.[23]

Importantly, it was demonstrated convincingly, that one or both of the loci encoding miR-101 were somatically lost in 37.5% of clinically localized prostate cancer cells and 66.7% of metastatic disease prostate cancer cells.[24]

Epigenetic aberrations may also cause alterations in miRNA expression. In fact ½ of the miRNA genes are associated with CpG islands and thus are candidates for regulation by DNA methylation.[25]

4. microRNA Profiling in Prostate Cancer

Alterations in miRNA expression have been observed in various human cancers including CaP. While changes in miRNA expression may not cause tumorigenesis, they may regulate genes important in tumor pathogenesis and may be useful as diagnostic and prognostic biomarkers for detecting cancers early, classifying tumors and predicting their outcome. miRNA expression profiles may also be useful for uncovering the etiologies of CaP and for discovering novel therapeutic targets for CaP. In the future, microRNA expression profiles may inform targeted drug therapies.

MicroRNAs may be the best biomarkers for cancer because of their involvement in development and their direct role in gene regulation. In contrast, only a small percentage of protein-coding mRNAs are regulatory molecules. Another advantage of microRNA expression profiling is the small number of microRNAs (695 human miRNAs as of Sanger miRBase version 12.0, September 2008.[26]) which will make it easier to implement in clinical setting.

5. Global Up or Down-Regulation of miRNAs in CaP?

There have been conflicting reports of global downregulation or upregulation of miRNA expression in the earliest studies of cancer vs. normal cells. In a microRNA expression analysis study involving a large number of tumors (540 samples across 6 tumor types including CaP), there was global upregulation of miRNA in all tumor types.[27] In a comparison of all tumor vs all normals, 26 miRNAs were overexpressed and 17 were underexpressed. For the prostate cancer samples (56 CaP and 7 normals from non-cancer individuals), 39 miRNAs are upregulated and 6 miRNAs are down-regulated.

In contrast, a separate study which also profiled miRNA expression in a large number (334 samples) of tumors (including 6 prostate adenocarcinoma samples and 8 normal prostates) showed exclusive global downregulation of miRNAs in tumors (129 of 131).[28] The authors of the latter study argue that the global downregulation of miRNA expression in tumors is consistent with the de-differentiation process found in cancers which parallels the de-differentiated embryonic state of organisms when few miRNAs are expressed. They also found that the least differentiated tumors had the lowest levels of miRNA expression. Many of the downregulated miRNAs are thought to have tumor suppressor function. However, Volinia et al. claim that they correctly identified proven oncogenic miRNAs such as miR-155, miR-17-5p, miR-92-1 and miR-21 whereas Lu et al. had not identified these miRNAs as being overexpressed. Rather, Lu et al. reported that miR-17-5p, miR-20 and miR-92 were down-regulated in tumors.

What could account for the discrepancies between the two studies? In the case of CaP the results from Lu et al. are based only on a very limited number (6 prostate adenocarcinomas and 8 normal prostates), and as such do not have the statistical power of the results from Volinia et al. (56 CaP and 7 normals from non-cancer individuals). One possibility for the downregulation of miRNAs in CaP would be if miRNA expression is higher in stromal cells and that the tumors have less stromal cells than the normal tissue.[29] In order to rule out this possibility the prostate tumors should be purified by laser capture microdissection (LCM). The microRNA profiling technologies used are also different.

First of all there were more miRNA probes in Volinia et al.'s study (245 miRNAs belonging to human and mouse[30]) than in Lu et al.'s paper (217 mammalian miRNAs). Importantly, the microarrays used by Volinia et al. included precursor miRNA probes in addition to mature miRNAs. Lu et al. were only observing the mature miRNAs since they gel-purified their total RNA.

Lu et al. used a bead-based solution hybridization method which they claim is more specific and accurate than the solid-phase microarray platform.

In a subsequent study, microRNA expression was profiled for 328 known and 152 novel human miRNAs in 10 benign peripheral zone tissues and 16 prostate cancer tissues (consisting of 70-90% cancer) using miRNA microarrays.[29] Similarly to Lu et al., the Baylor College of Medicine study conducted by Ozen et al. found universal downregulation of miRNAs (76 of the 85 detected miRNAs, P<0.05, t-test) in clinically localized CaP relative to benign peripheral zone tissue.

Mattie et al. also found predominantly downregulated miRNA expression in CaP on two biopsies of cancer vs. pooled normal tissue using microarray-based approach.[31] Porkka et al. analyzed miRNA expression profiling of 4 benign prostatic hyperplasia (BPH) and 9 CaP samples and found 36 downregulated miRNAs (22 were downregulated in all carcinoma samples (untreated CaP and hormone refractory prostate cancers) and 15 were down-regulated only in HRPC) and 14 upregulated miRNAs (8 increased miRNA expression in all carcinomas and 6 of them only in HRPC).[32] Ozen et al. and Mattie et al. had used purified mature miRNA just as Lu et al. had done, whereas Volinia et al. labeled the total RNA for the hybridization.

Thus, Volinia et al. were able to detect pre-miRNAs and miRNAs since they had probes for the pre-miRNA and miRNA on their arrays. If there was a block in pre-miRNA processing in CaP without a corresponding decrease in transcription, this could explain the difference in global miRNA upregulation of downregulation between Volinia et al.'s study and the other three studies. Interestingly, Porkka et al. purified their total RNA using columns with a molecular weight cut-off of 300 nucleotides prior to labeling and hybridizing the RNAs. With this purification scheme, they should have been able to observe the binding of the pre-miRNAs as well and they report 50 differentially expressed miRNAs between CaP and BPH 36 (72%) down-regulated and (28%) upregulated. Thus, Porkka et al. observed a greater percentage of upregulation than in Ozen et al. or Lu et al. or Mattie et al.'s studies perhaps because pre-miRNA were included in their microarray hybridization.

Another source of discrepancy is in the nature of the tissues which were analyzed. Ozen et al. used benign peripheral zone tissue from radical prostatectomy samples as the normals while Volinia et al. used normal from non-cancer individuals. Furthermore, the cancer cases analyzed by Ozen et al. are all from localized CaP while the clinical and pathological stage of the CaP used by Volinia et al. are not stated.[29] Finally, the method of tissue harvesting may significantly impact the measured microRNA expression levels. It has been shown that hypoxia[33] and other cell stresses[34] can impact miRNA expression.

6. Differentially Expressed miRNAs between CaP and Normal Cells

A list of differentially expressed miRNAs between CaP and normal cells is shown in **Table 1** (microRNAs down-regulated in CaP) and in **Table 2** (microRNAs up-regulated in CaP). While there are many differences, there are some overlaps in the miRNAs found across studies which are indicated in **bold** font.

Let-7c is increased in CaP. One of the evidence for this is by LNA (locked nucleic acid) hybridization.[35] The authors used a novel nucleotide the LNA-mTP (locked methyl cytidine 5' triphosphate) to transcribe short (30 nt) LNA-RNA probes with which they performed dot blot hybridization on nylon membranes. When they tested prostate cancer tissues, they were able to detect let-7c. While the LNA/RNA hybridization may be more specific than DNA/RNA hybridization, and thus distinguish between the closely related members of the let-7 family, the miRNA microarray used by Volinia et al. was able to detect the increased expression of let-7d and let-7i by PAM (Predicted Analysis of Microarrays) scores.

In direct contrast to these findings, Ozen et al. reported the downregulation in the expression of miRNAs in the let-7 family (7b-g, 7i). Ozen et al. validated the downregulation of let-7c in prostate cancer cells using RT-PCR assays. The differences in these observations may be that Ozen et al. have used the benign prostate peripheral zone tissues as the normals to compare the miRNA expression with the prostate cancer cells. The benign prostate cells may be already undergoing changes in their miRNA expression profile, and as such the expression of let-7c may indeed be downregulated in comparison to cells obtained from prostate cancer tissue.

Likewise, Porkka et al. also reported a down regulation of let-7a, b, d, f, g RNAs in 9 CaP samples vs. 4 BPH samples and they had used benign prostate hyperplasia samples just as Ozen et al. had. Surprisingly, Volinia et al. also reported a downregulation of the let-7a just as Porkka et al. described. It is not clear whether Volinia et al.'s microarrays can distinguish precisely between let-7a which was upregulated and let-7d and let-7i which were downregulated in the same CaP samples used in their study. Porkka et al. did not report any upregulated let-7 microRNAs.

Table 1. microRNAs differentially expressed in CaP vs. normal: miRNAs down-regulated in CaP.

Ozen et al.	Ambs et al.	Esqet Jay et al.	Volinia et al.	Tong et al.	Varambally et al.	Mattie et al.	Porkka et al.
let-7 family (7b-g, 7i)	miR-520h	**mir-15a**	let-7a	mir-23b	mir-101 (targets EZH2)	**mir-15a**	Let-7a,b,c,d,g.
miR-26-b	miR-494	**mir-16**	miR-24	**mir-145**		mir-126	**miR-16**
miR-29a-c	miR-490		**miR-29a**			mir-143	miR-23a
miR-30a-e	miR-1(-2)		miR-128a				miR-23b
miR-99a-b	miR-133a -1		miR-149				miR-26a
miR-**125a-b**	mir-218-2		miR-218				miR-92
miR-200a-b	miR-220						miR-99a
miR-145	miR-128a						miR-103
miR-205	miR-221						**miR-125a**
	miR-499						**miR-125b**
	miR-329						miR-143
	miR-340						**miR-145**
	miR-345						miR-195
	miR-410						miR-199a
	miR-126						**miR-221**
	miR-205						**miR-222**
	miR-7-1/2						miR-497
	miR-145						**let-7f**
	miR-34a						miR-19b
	miR-487						miR-22
	let-7b						miR-26b
							miR-29a
							miR-29b
							miR-30a-5p
							miR-30b
							miR-30c
							miR-100
							miR-141
							miR-148a,
							miR-205

Table 2. microRNAs differentially expressed in CaP vs. normal: miRNAs upregulated in CaP.

Volinia et al.	Ambs et al.	Galardi et al Tong et al.	Shi et al.	Porkka et al.	Lee et al.
miR-21	miR-32	mir-221	**miR-125b**	miR-202	**miR-125b**
miR-191	miR-182	miR-222	miR-92	miR-210	
miR-17-5p	miR-31		miR-191	mir-296	
let-7d	miR-26a		**miR-106a**	miR-320	
let-7i	miR-125a		mir-145	miR-370	
miR-16	miR106b/miR-93/		miR-21	miR-373	
miR-20a	miR-25		let-7c	miR-498	
miR-25	miR 99b cluster			miR-503	
miR-26a	(miR-99b/miR-125a)			**miR-184**	
miR-27a				**miR-198**	
miR-29a				miR-302	
miR-29b				miR-345	
miR-30c				miR-491	
miR-32				miR-513	
miR-34a					
miR-92					
miR-93					
miR-95					
miR-101					
miR-106a					
miR-124a					
miR-135					
miR-146					
miR-148					
miR-181b					
miR-184					
miR-187					
miR-191					
miR195					
miR-196					
miR-197					
miR-198					
miR-199					
miR-203					
miR206					
miR-214					
miR-223					

Shi et al. reported the increased expression of miR-125b, miR-92, miR-191, miR-106a, mir-145, miR-21 in CaP cell lines relative to non-malignant prostate cell lines using Northern blotting analysis.[36] Volinia et al. had also reported upregulation of miR-21, miR-92, miR-191 and miR-106a in CaP solid tumors.

In contradiction to this finding, Ozen et al. saw a decrease in miR-125b using both miRNA microarray and RT-PCR assays. In agreement with Ozen et al., Porkka et al. also reported a down-regulation of miR-125b in prostate carcinoma sample compared to BPH. However, the difference between the two studies and that of Shi et al. is that the latter group compared miRNA expression between prostate cancer cell line and benign prostate epithelial cell lines whereas Ozen et al. and Porkka et al. compared miRNA expression between prostate tumors and benign peripheral prostate zone tissue.

Shi et al., however, also confirmed the detection of miR-125b in prostate tissues using LNA-ISH (locked nucleic acid-in situ hybridization) and reported that miR-125 b is more highly expressed in 10 primary CaP samples with Gleason scores of 6 to 8 than in 2 benign prostatic tissues.

In a study of 60 primary prostate tumors and 16 nontumor prostate tissues, Ambs et al. report the following microRNA gene signature of prostate cancer. The following microRNAs were highly downregulated: miR-520h, miR-494, miR-490, miR-1(-2) and miR-133a -1. By microRNA array measurements, the fold changes were modest ranging from 0.3 fold in miR-520h to 0.6-fold in miR-1(-3).

The upregulated microRNAs were: miR-32, miR-182, miR-31, miR-26a. Again, the fold-changes in microRNA expression between the prostate tumor and the nontumor prostate tissue were not dramatic: miR-21 which was most significantly upregulated was only expressed 2.1-fold higher in the tumor vs. nontumor. An interesting group of microRNAs that were upregulated 1.4-fold in the tumors was the miR-106b-25 cluster (miR-106b/miR-93/miR-25).

Ambs et al. had carried out the simultaneous profiling of the mRNA on the same samples which were profiled for microRNA expression. Hence they were able to confirm the co-expression of some of the microRNAs with their host mRNA genes. For instance, the upregulation of both miR-32 and the miR-106b-25 cluster is consistent with the increased expression of their respective host genes C9orf5 (2.1-fold up-regulation) and MCM7 (1.7-fold up regulation) in prostate tumors.

The differentially expressed microRNAs discovered by microarray were confirmed using RT-PCR on a random subset of the tumors and non-tumor tissues. The RT-PCR results showed greater fold-changes than those obtained by the microarray. The miR-32 (average 3.2 fold) and miR-106 (average 3.0 fold) were upregulated in tumors compared to non-tumor tissues. miR-1 was downregulated (avg., 0.44-fold) consistent with the microarray results.

7. Differentially Expressed miRNAs between AI and AD Prostate Cancer Cells

Jiang et al. observed a 4-fold increase of miR-100 in AI prostate cancer cell lines and a 53-fold increase in let-7c in AD LnCaP cell lines compared to control cell lines using RT-PCR assay.[37] Interestingly, this RT-PCR assay analyzed the expression of precursor miRNA and not mature miRNA. Notice the predominant upregulation of miRNAs in CaP when the precursor miRNA is analyzed. This is in agreement with the results of Volinia et al. who were also measuring pre-miRNA and mature miRNA.

Shi et al., Jiang et al. and Mattie et al. also observed that miR-100 is upregulated in AI vs. AD CaP cells. Using primer extension analysis, Lee et al. observed high expression of miR-125b in PC3 cells which are AI CaP cells.[37] In a direct comparison of AI to AD CaP cell lines, Lin et al. found using microarray service from LCS Sciences that three miRNAs (miR-184, miR-361 and miR-424) were upregulated whereas eight miRNAs (miR-19b, miR-29b, miR-128b, miR-146a, miR-146b, miR-221, miR-222, and miR-663) were down-regulated.

Significantly, Porkka et al. had also found miR-184 to be upregulated only in AI clinical HRPC tumors and that miR-19b and miR-29b were down-regulated only in AI clinical HRPC patient samples. Lin et al. selectively validated the increase of miR-184 and the loss of miR-146 in human prostate cancer tissue arrays.

There are other overlaps between miRNAs differentially expressed in AI vs. AD CaP cells that we can see in **Table 3, 4**. For example, miR-100 is upregulated in AI vs. AD CaP cells in by Shi et al., Jiang et al. and Mattie et al. Both Shi et al. and Mattie et al. observed upregulation of miR-30c in AI vs AD CaP. However, there were some marked differences as well. For example, Mattie et al. found miR-221 and miR-222 are up-regulated in AI vs. AD cells whereas Lin et al. found miR-221 and miR-222 to be downregulated. We must assume that the types of cell lines Lin et al. used to compare AI vs. AD CaP are more accurate measure of differences in androgen responsive than the types of cell lines Mattie et al. used since Lin et al. used basically the AI cell lines that were derived from the same corresponding AD cell lines. That is, Lin et al. used LNCaP vs. LNCaP C4-2B and PC3 vs PC3-AR9 while Mattie et al. used LnCaP vs. PC3.

Table 3. microRNAs differentially expressed between AI and AD CaP cells. miRNAs upregulated in AI vs. AD.

Jiang et al.	Mattie et al.	Shi et al.	Lin et al.	Lee et al.	Porkka et al.
miR-100	**miR-100**	**miR-125b**	**miR-184**	**miR-125b**	**miR-184**
	miR-30c	miR-16	mir-361		miR-198
	miR-21	**miR-21**	mir-424		miR-302c
	miR-221	miR-30c			miR-345
	miR-222	**miR-100**			miR-491
					miR-513

Table 4. microRNAs differentially expressed between AI and AD CaP cells: miRNAs downregulated in AI vs. AD.

Jiang et al.	Mattie et al.	Shi et al.	Lin et al.	Porkka et al.
let-7c	miR-99a	miR-92 miR-106a	miR-19b	let-7f
	miR-200c		**miR-29b**	miR-19b
	miR-148a		miR-128b	miR-22
	miR-34a		miR-146a	miR-26b
	let7c		miR-146b	miR-27a
	let7b		miR-221	miR-27b
	miR106b		miR-222	miR-29a
			miR-663	**miR-29b**
				miR-30a-5p
				miR-30b
				miR-30c
				miR-100
				miR-141
				miR-148a
				miR-205

8. Functions of miRNAs Differentially Expressed in CaP

There has been very limited investigation of the functions of the miRNAs which are differentially expressed in CaP and the validation of their targets. Overexpression of miR-125b using chemically modified synthetic miR-125b in three AD CaP cell lines stimulated the growth of these cell lines in the absence of androgens and caused a 3-fold increase in the S-phase fraction in LNCaP cells compared to controls.[36] On the other hand, when miR-125b activity is repressed in AI LNCaP-cds cells by transfection with anti-miR-125b, cell growth was inhibited and there was 54% reduction of the S-phase fraction and an increase in sub-G1 cells relative to controls. These functional studies demonstrated that miR-125b is involved in AI growth and CaP survival.

Similar knock-down and knock-in studies in cell lines showed that miR-20a (overexpressed in clinical CaP cells as shown by Volinia *et al.*) was oncogenic via its anti-apoptotic function in PC3 CaP cells.[38] Ectopic expression of miR-146a in PC3 cells which do not have this microRNA reduces their proliferation rate.[39] This suggest that miR-146a is a tumor suppressor that inhibits PC3 cell growth. Ectopic expression of miR-221 and or miR-222 induced a >2-fold increase in growth rate and a 3-4-fold increase in the colony formation in miRNA-transfected LNCaP cells. Inhibition of miR-221/222 with antisense oligonucleotides in PC3 cells reduce 4.5 fold the number of colonies growing in soft agar.

Thus, miR-221 and miR-222 may contribute to the growth and progression of CaP. The ectopic expression of miR-126 resulted in a reduction in migration and invasiveness of LNCaP cells which supports the role of miR-126 in the malignant phenotype of CaP cells.[40]

With the invention of the high-throughput Dharmacon miRNA functional screening assay, it is expected that many more miRNAs will be designated a function. The assay can knock-down or knock-in every miRNA in a cell whereupon the effects of these perturbations are monitored with a functional assay in order to discover the microRNAs that are involved in a particular function.

9. Target Validation

Very few targets of the miRNAs differentially expressed in CaP have been selected for validation as summarized in **Table 5, 6**. E2F1 promotes apoptosis in PC3 cells, and is a target of miR-20a. Interestingly, E2F1 regulates the expression of miR-20a by binding to the promoter of the miR-17-92 cluster which includes the miR-20a. Thus, there is an autoregulatory feedback loop between E2F1 and miR-20a.[38] Bak1, a target of miR-125b, is a member of the Bcl-2 family and functions a pro-apoptotic regulator by opposing the functions of the anti-apoptotic Bcl-2 family proteins. $p27^{Kip1}$, a target of miR-221/miR-222 is a cyclin-dependent kinase inhibitor.[3]

Table 5. miRNA targets in CaP : Pro-apoptotic or growth inhibitory genes.

Down	Up
E2F1 (Sylvestre et al.) growth inhibitory/proapoptotic in PC3 cells	miR-20a
Bak1 (deVere White) proapoptotic **SPGL1** (Bandhuvula and Saba) tumor suppressor and cancer surveillance	miR-125b
RB1 (Volinia et al.)	miR-106a
P27 Kip1 (Galardi et al.) cell cycle inhibition	miR-221/222
EZH2 (Varambally et al.) histone methyltransferase	miR-101

Table 6. miRNA targets in CaP: Proliferation-related genes.

Up	Down
SLC45A3 (PCANAP6) (Musiyenko et al.) determinant of mobility and invasiveness of CaP cells	miR-126*
ROCK1 (Lin et al.) involved in AI growth and metastasis of PC3 cells	miR-146a
E2F3 (Ozen et al.)	miR-34a
RAS (Ozen et al.)	Let-7
MCL-1 (Ozen et al.)	miR-29a
BCL2 (Ozen et al.)	miR-16

Overexpression of miR-221/miR-222 induces a G1 to S shift in cell cycle and is able to induce a powerful enhancement of their colony-forming potential in soft agar. Conversely, a knock-down of miR-221/miR-222 through antisense oligonucleotides increases p27^{Kip1} in PC3 cells and reduces their clonogenicity in vitro. Thus, miR-221/miR-222 is an oncomir which contributes to ontogenesis and CaP progression by downregulating the tumor suppressor p27^{Kip1}. ROCK1, a target of miR-146a is a Rho-activated protein kinase gene that is involved in AI growth and metastasis of PC3 cells.[41] EZH2, a target of miR-101, is a histone methyltransferase that contributes to the epigenetic silencing of target genes and that regulates the survival and metastasis of cancer cells.[24]

The methods for target validation include bioinformatic prediction using Targetscan, PicTar, miRanda followed by experimental validation including reporter assays, Western Blot and transient transfection.[42] Clearly, a more high-throughput genomic technology such as microarrays or SILAC experiments[43] should be used to discover the multiple functional targets that a single microRNA can have.

10. Regulation of miRNAs by Androgen

Since the growth of prostate cells is regulated by androgens, it is plausible that miRNA expression is regulated by the Androgen Receptor (AR). Recently, evidence has been accumulating for the regulation of miRNA expression by AR. Porkka et al. reported that miRNA expression changed depending on the androgen receptor status (+ or -) in CaP cell line and xenografts, suggesting that androgens may regulate the expression of miRNAs.[32] Similarly, Lin et al. reported that the transfection of AR into PC3 cells induced the differential expression of 11 miRNAs compared to AR-negative PC3 cells.[39]

Shi et al. observed direct evidence that androgen-AR signaling regulates the expression of a subset of miRNAs. In the absence of androgens, reduced levels of miR-125b were detected by LNA-ISH in AR-positive CaP cell lines. Androgen treatment stimulated an increase in miR-125b.[44] Furthermore, transfection of AR into AR-negative prostate epithelial cells followed by treatment with androgen also increased the expression of miR-125b. Chromatin IP analysis showed that AR was recruited to the 5' DNA region of the miR-125b-2 locus in an androgen-dependent manner.

Ambs et al.[45] found that treatment of the androgen-insensitive DU145 prostate cell lines did not result in any changes in microRNA expression, whereas treatment of the androgen sensitive LNCaP cells changed after R1881 treatment. One of the microRNAs miR-338 was significantly upregulated whereas the other microRNAs were down-regulated (miR-126-5p, miR-146b, miR-219-5p and all members of the miR-181b-1, miR-181c and miR-221 clusters).

A comparison of the baseline microRNA expression in androgen-insensitive DU145 cells and androgen-dependent LnCaP cells showed that all members of the three microRNA clusters were significantly overexpression in the AI DU145 cells than in the AD LnCAP cells. Interestingly, these microRNA clusters all have putative androgen receptor binding sites in their flanking regions.[45]

11. Regulation of miRNA in CaP by Dicer and Other Factors

Ambs et al.[45] profiled mRNA expression in the same prostate cancer samples that were profiled for microRNA expression. They found that the microRNA processing proteins Dicer and DGCR8 were upregulated in CaP tumors. This confirms a similar report by[46] and may suggest that prostate tumors process pre-miRNAs to mature miRNAs more efficiently than normal prostate tissue. However, Dicer may have functions independent of microRNA processing in prostate tumors.

The upregulation of Dicer may also be associated with the increase of miRNA expression in some of the CaP miRNA profiling studies. Since microRNA processing enzymes are general factors, additional specific factors may be required for regulation of individual miRNAs. For instance the tumor suppressor p53 activates miR-34a in CaP.[47]

12. Conclusion

While there are some successes with identifying microRNAs that are differentially expressed in CaP and in determining their function and targets, the genomic profiling of microRNA expression is fraught with problems. First, the technologies used to profile the global expression of miRNAs are not very comparable because there is no standard method of sample collection, RNA purification, RNA labeling and hybridization. Clearly, there are differences in miRNA expression depending on whether the mature miRNA have been purified away from the precursor miRNAs.

The microarray platforms which analyze both the mature miRNA and precursor miRNA expression seem to show a global upregulation of miRNA expression whereas platforms that analyze only the purified mature miRNA exhibit a global downregulation of miRNA expression.

The miRNA profiling technologies must be standardized. A multi-center, cross-platform study similar to the Microarray Quality Control (MAQC) Consortium must be done on the same standardized samples in order to facilitate the comparison of biological results across studies. Such an undertaking is on the way through the efforts of the ABRF (Association of Biomolecular Resource Facilities) Microarray Research Group.

Another technical issue is that the microarray platforms may not have sufficient specificity to distinguish paralogous microRNAs which have very similar sequences. One solution to this problem would be to use next-generation deep sequencing to profile the expression of microRNA samples.

In addition, the high-throughput RT-PCR based microRNA profiling technologies such as the ABI Taqman miRNA arrays may prove to be a more accurate measure of microRNA expression.

It is difficult to make generalizations about the role of microRNAs in prostate cancer across studies, because the prostate cancer samples are simply not comparable. Different cell lines are used to represent AI and AD CaP cells.

The definitions of control normal prostate cells are also different among studies, some using BPH and others using non-cancer prostate cells; the genomic profiling will be very different depending on which types of cells are chosen as the control to compare with the CaP. This is particularly so in CaP where the tumor is likely accompanied by field changes in the prostate that affect the normal epithelium.

Gene expression profiling of normal prostate tissues harvested from organ donors in the absence of prostate cancer show very different gene expression profiles from normal prostate tissues harvested from organs containing prostate cancers.[46]

A truly useful CaP biomarker must based on the amount of Gleason histological grade 4/5 because for every 10% increase in Gleason grade 4/5 in the largest peripheral zone of the prostate, we lose 10% of cures by radical prostastectomy.[5] Gleason 1, 2, 3 never result in cancer. It would be useful to focus on Gleason 4/5 grade CaP for the purpose of biomarker discovery. Sample sizes used in the studies thus far are too small in, and clearly, larger numbers of CaP samples must be analyzed.

Histologically, prostate cancer tends to infiltrate the normal.[48] Therefore most prostate cancer samples will contain contaminating normal prostate glands as well as stromal cells, thus making the samples very heterogeneous. Perhaps LCM purification of the tumor cells from stromal cells may reduce the discrepancies between studies.

The discovery of targets cannot be continued painstakingly, one target at a time. This makes it difficult to find but single targets for a handful of arbitrarily selected microRNAs. Target discovery should be done with genomic technologies such as the combination of a miRNA knock-down or knock-in and microarrays or SILAC type of proteomic approach.[43,49] Likewise, the discovery of microRNA function will be accelerated through the use of high-throughput miRNA functional screens such as Dharmacon's MIRIDIAN microRNA inhibitor and hairpin libraries.

To conclude, it appears that microRNAs are clearly involved in prostate cancer either as causative agents regulating the targets or responsive elements that are regulated by other factors. With more standardized methods of microRNA expression profiling analysis and better characterization of larger clinical sample datasets, it is likely that important molecular diagnostics and prognostic markers for prostate cancer will be discovered. Finally, the elucidation of the roles microRNAs play in prostate cancer may lead to improved therapeutics for prostate cancer.

References

1. Jemal, A. et al. Annual report to the nation on the status of cancer, 1975-2005, featuring trends in lung cancer, tobacco use, and tobacco control. J Natl Cancer Inst 100, 1672-94 (2008).
2. Sheldon, C. A., Williams, R. D. & Fraley, E. E. Incidental carcinoma of the prostate: a review of the literature and critical reappraisal of classification. J Urol 124, 626-31 (1980).
3. Galardi, S. et al. miR-221 and miR-222 expression affects the proliferation potential of human prostate carcinoma cell lines by targeting p27Kip1. J Biol Chem 282, 23716-24 (2007).
4. Scardino, P. Update: NCCN prostate cancer Clinical Practice Guidelines. J Natl Compr Canc Netw 3 Suppl 1, S29-33 (2005).
5. Kirby, R. S. Prostate Cancer: principles and practice (Taylor and Francis, London, New York, 2006).
6. Li, S., Bhamre, S., Lapointe, J., Pollack, J. R. & Brooks, J. D. Application of genomic technologies to human prostate cancer. Omics 10, 261-75 (2006).
7. McNeal, J. E. & Gleason, D. F. [Gleason's classification of prostatic adenocarcinomas]. Ann Pathol 11, 163-8 (1991).
8. Oh, W. K. & Kantoff, P. W. Management of hormone refractory prostate cancer: current standards and future prospects. J Urol 160, 1220-9 (1998).
9. Lewis, B. P., Burge, C. B. & Bartel, D. P. Conserved seed pairing, often flanked by adenosines, indicates that thousands of human genes are microRNA targets. Cell 120, 15-20 (2005).
10. Borchert, G. M., Lanier, W. & Davidson, B. L. RNA polymerase III transcribes human microRNAs. Nat Struct Mol Biol 13, 1097-101 (2006).
11. Lee, Y. et al. MicroRNA genes are transcribed by RNA polymerase II. Embo J 23, 4051-60 (2004).
12. Denli, A. M., Tops, B. B., Plasterk, R. H., Ketting, R. F. & Hannon, G. J. Processing of primary microRNAs by the Microprocessor complex. Nature 432, 231-5 (2004).
13. Gregory, R. I. et al. The Microprocessor complex mediates the genesis of microRNAs. Nature 432, 235-40 (2004).
14. Lee, Y. et al. The nuclear RNase III Drosha initiates microRNA processing. Nature 425, 415-9 (2003).
15. Tong, A. W. et al. MicroRNA profile analysis of human prostate cancers. Cancer Gene Ther (2008).
16. Hutvagner, G. et al. A cellular function for the RNA-interference enzyme Dicer in the maturation of the let-7 small temporal RNA. Science 293, 834-8 (2001).
17. Salzman, D. W., Shubert-Coleman, J. & Furneaux, H. P68 RNA helicase unwinds the human let-7 microRNA precursor duplex and is required for let-7-directed silencing of gene expression. J Biol Chem 282, 32773-9 (2007).

18. Filipowicz, W., Bhattacharyya, S. N. & Sonenberg, N. Mechanisms of post-transcriptional regulation by microRNAs: are the answers in sight? Nat Rev Genet 9, 102-14 (2008).
19. Calin, G. A. et al. Human microRNA genes are frequently located at fragile sites and genomic regions involved in cancers. Proc Natl Acad Sci U S A 101, 2999-3004 (2004).
20. Ambros, V. The evolution of our thinking about microRNAs. Nat Med 14, 1036-40 (2008).
21. Calin, G. A. et al. MicroRNA profiling reveals distinct signatures in B cell chronic lymphocytic leukemias. Proc Natl Acad Sci U S A 101, 11755-60 (2004).
22. Yeager, M. et al. Genome-wide association study of prostate cancer identifies a second risk locus at 8q24. Nat Genet 39, 645-9 (2007).
23. Glinsky, G. V. An SNP-guided microRNA map of fifteen common human disorders identifies a consensus disease phenocode aiming at principal components of the nuclear import pathway. Cell Cycle 7, 2570-83 (2008).
24. Varambally, S. et al. Genomic Loss of microRNA-101 Leads to Overexpression of Histone Methyltransferase EZH2 in Cancer. Science (2008).
25. Brueckner, B. et al. The human let-7a-3 locus contains an epigenetically regulated microRNA gene with oncogenic function. Cancer Res 67, 1419-23 (2007).
26. http://microrna.sanger.ac.uk/cgi-bin/sequences/browse.pl
27. Volinia, S. et al. A microRNA expression signature of human solid tumors defines cancer gene targets. Proc Natl Acad Sci U S A 103, 2257-61 (2006).
28. Lu, J. et al. MicroRNA expression profiles classify human cancers. Nature 435, 834-8 (2005).
29. Ozen, M., Creighton, C. J., Ozdemir, M. & Ittmann, M. Widespread deregulation of microRNA expression in human prostate cancer. Oncogene 27, 1788-93 (2008).
30. Liu, C. G. et al. An oligonucleotide microchip for genome-wide microRNA profiling in human and mouse tissues. Proc Natl Acad Sci U S A 101, 9740-4 (2004).
31. Mattie, M. D. et al. Optimized high-throughput microRNA expression profiling provides novel biomarker assessment of clinical prostate and breast cancer biopsies. Mol Cancer 5, 24 (2006).
32. Porkka, K. P. et al. MicroRNA expression profiling in prostate cancer. Cancer Res 67, 6130-5 (2007).
33. Kulshreshtha, R. et al. A microRNA signature of hypoxia. Mol Cell Biol 27, 1859-67 (2007).
34. Marsit, C. J., Eddy, K. & Kelsey, K. T. MicroRNA responses to cellular stress. Cancer Res 66, 10843-8 (2006).
35. Kore, A. R., Hodeib, M. & Hu, Z. Chemical Synthesis of LNA-mCTP and its application for MicroRNA detection. Nucleosides Nucleotides Nucleic Acids 27, 1-17 (2008).
36. Shi, X. B. et al. An androgen-regulated miRNA suppresses Bak1 expression and induces androgen-independent growth of prostate cancer cells. Proc Natl Acad Sci U S A 104, 19983-8 (2007).
37. Jiang, J., Lee, E. J., Gusev, Y. & Schmittgen, T. D. Real-time expression profiling of microRNA precursors in human cancer cell lines. Nucleic Acids Res 33, 5394-403 (2005).
38. Sylvestre, Y. et al. An E2F/miR-20a autoregulatory feedback loop. J Biol Chem 282, 2135-43 (2007).
39. Lin, S. L., Chiang, A., Chang, D. & Ying, S. Y. Loss of mir-146a function in hormone-refractory prostate cancer. Rna 14, 417-24 (2008).

40. Musiyenko, A., Bitko, V. & Barik, S. Ectopic expression of miR-126*, an intronic product of the vascular endothelial EGF-like 7 gene, regulates prostein translation and invasiveness of prostate cancer LNCaP cells. J Mol Med 86, 313-22 (2008).
41. Lin, S. L., Chang, D. & Ying, S. Y. Hyaluronan stimulates transformation of androgen-independent prostate cancer. Carcinogenesis 28, 310-20 (2007).
42. Kuhn, D. E. et al. Experimental validation of miRNA targets. Methods 44, 47-54 (2008).
43. Baek, D. et al. The impact of microRNAs on protein output. Nature 455, 64-71 (2008).
44. Shi, X. B., Tepper, C. G. & White, R. W. MicroRNAs and prostate cancer. J Cell Mol Med 12, 1456-65 (2008).
45. Ambs, S. et al. Genomic profiling of microRNA and messenger RNA reveals deregulated microRNA expression in prostate cancer. Cancer Res 68, 6162-70 (2008).
46. Chiosea, S. et al. Up-regulation of dicer, a component of the MicroRNA machinery, in prostate adenocarcinoma. Am J Pathol 169, 1812-20 (2006).
47. Rokhlin, O. W. et al. MicroRNA-34 mediates AR-dependent p53-induced apoptosis in prostate cancer. Cancer Biol Ther 7, 1288-96 (2008).
48. Ruijter, E. T., van de Kaa, C. A., Schalken, J. A., Debruyne, F. M. & Ruiter, D. J. Histological grade heterogeneity in multifocal prostate cancer. Biological and clinical implications. J Pathol 180, 295-9 (1996).
49. Selbach, M. et al. Widespread changes in protein synthesis induced by microRNAs. Nature 455, 58-63 (2008).

CHAPTER 6

INTRONIC MICRORNA: CREATION, EVOLUTION AND REGULATION

Sailen Barik[1] and Titus Barik[2]

[1] *Department of Biochemistry and Molecular Biology, University of South Alabama, College of Medicine, 307 University Boulevard, Mobile, Alabama 36688-0002, USA*
[2] *770 Soaring Drive, Marietta, GA 30062*

MicroRNAs (miRNAs or miRs) are short non-coding RNAs that down-regulate gene expression by silencing specific target mRNAs through translational inhibition and/or degradation. Interestingly, a large fraction of miRNAs maps within introns of their "host" genes, generating potentially interesting regulatory circuits, the full significance of which has barely been appreciated. In this review, we offer a comprehensive yet short treatise on the nature, location, evolution, regulation and potential mechanisms of action of the different types of intronically situated miRNAs. It is hoped that the outline presented here will provoke novel thoughts and research in this new and exciting miRNA world.

1. Introduction: miRNAs Trigger RNAi

RNA interference (RNAi) is a physiological pathway in which miRNAs silence specific target RNAs.[1-3] The miRNAs are themselves processed from larger precursors (called Primary miRNA or pri-miRNA) through multiple steps that are still not fully clear.[4,5] Processing of a hairpin precursor can in theory produce two miRNAs from the two strands, denoted by suffix -5p or -3p (e.g., miR-126-5p and miR-126-3p).[6] For unknown reasons, the two products are often found in highly unequal amounts such that one of them predominates. For example, miR-126-3p can be as much as 100-times more abundant than its fraternal twin, miR-126-5p.[7,8]

The miRNAs recognize the miRNA-response elements (MREs), short RNA sequences located in the 3' UTR of the target mRNAs; the recognition is coupled with the assembly of a multi-protein complex at the site, known as of the RNAi-induced silencing complex (RISC).[9-13] The rules of miR-MRE pairing are only partly understood. A near-perfect match with nucleotides 2-8 at the 5' end of the miRNA, known as the "seed" (or "core") sequence, is minimally required, but other features are clearly needed that remain a mystery. The most cardinal component of RISC is a member of the Argonaute (Ago) family of PAZ-PIWI domain proteins.[14] Four Ago paralogs (Ago1-4) are present in humans; of these, Ago2 is unique because of its pronounced RNase activity that cleaves the target mRNA within the miR-MRE paired region.[15,16] Thus, Ago2-containing RISCs can cleave the target mRNA, whereas an Ago1-RISC instead blocks translation initiation or elongation of the mRNA.[17-20]

In either event, the net result is abrogation of protein synthesis, leading to silencing. Intriguingly, it is unclear what molecular rules dictate the preferential recruitment of a specific Ago. The emerging mechanism suggests that a perfect or near-perfect miR-MRE pairing recruits Ago2 whereas a relatively imperfect pairing recruits Ago1.[21] This could be due to structural differences in the RNA-binding domains of these Ago proteins or a RISC subunit that recognizes the pairing. Regardless of the mechanism, it is reasonable to postulate that there is no mathematically precise switch-over point and that a semi-perfect miR-MRE base-pairing may recruit either Argonaute. This is easily testable.

Roughly 500 human miRNAs have been experimentally validated,[6,7] and together they appear to regulate more than a third of all genes, impacting on a large variety of biological processes including development and differentiation, genome stability and imprinting, aging, oncogenesis and host-pathogen interaction. An ever-growing list of malfunctions of these processes is correlated to an aberrant miRNA expression or function.[22-26] Consistent with the occurrence of RNAi in eukaryotes, a disproportionately high number of MREs are found in genes unique to a multicellular body design, such as nerve development, signal transduction, and cell-cell communication, whereas few are found in genes involved in intermediary metabolism that is common to all cells.[22,27] Clearly, a greater understanding of miRNA biogenesis and function is essential to complete our knowledge of mammalian biology.

2. Intronic miRNA: A Classification

Interestingly, about half of all miRNAs map within the introns of protein-coding "host" genes, many of which are conserved across species, assuring significance.[8,27-32] However, the regulation of the "intron-derived" (or "intronic") miRNAs and their relationship with their host genes has remained an unsolved puzzle. We would like to classify the intronic miRNAs into two fundamentally different kinds (Fig. 1).

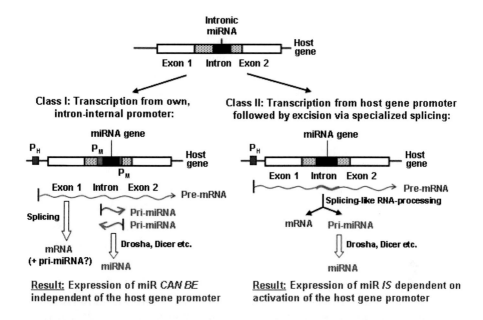

Fig. 1. Classification of Intronic miRNA genes. All intronic miRNA genes reside (Black box) inside an intron (Speckled box) of its "host" gene. In Class I, the miRNA gene has its own promoter (P_M, in Green), which can be in either orientation. Two types of precursor transcripts, the full-length pre-mRNA (Red) and the miRNA precursors (Pre-miRNA, in Green) are transcribed. The Pre-mRNA is processed by standard RNA splicing machinery to generate mRNA that is translated to protein. It is unknown whether the intronic pri-miRNA hairpin in the pre-mRNA is also processed by an RNase III activity to generate more pri-miRNA. In Class II, the host promoter drives all transcription, and the single primary transcript produces both mRNA and miRNA through RNA processing. The contrasting results are shown at the bottom (Blue).

2.1. Class I intronic miRNA

These are actually independent genes with their own dedicated promoters that just happen to be located within an intron of another gene.[33] Barring a few that uses Pol III promoters[34], these miRNA genes are transcribed from Pol II promoters (P_M in Fig. 1).[35,36] Thus, the majority has the features of a standard eukaryotic gene, including a dedicated Pol II promoter, enhancers, and a terminator. Most pri-miRNAs in fact contain 5'-cap and 3'-poly (A) tail.[35,36] Clearly, these miRNA genes will be activated or repressed by transcription factors and chromatin remodeling in response of environmental triggers. Unfortunately, very few miRNA genes, intronic or otherwise have been characterized in any detail.

In a human genome-wide study, proximal promoters of 175 human miRNAs were characterized by combining nucleosome mapping with chromatin signatures for promoters.[37] Roughly one-third of intronic miRNA genes were found to have promoters independent of their host gene promoters (the remaining likely share the host gene promoter, and therefore, belong to Class II; see later). At least 9 were regulated by the MITF transcription factor/oncoprotein in melanoma cells.

Similar comprehensive analysis of the C. elegans genome revealed over 100 miRNAs.[38] In an attempt to characterize their promoters, transgenic C. elegans strains were created, each expressing GFP under the control of putative miRNA promoter sequences. GFP expression was detected in a variety of tissues, although expression occurred later in development, consistent with functions after initial body-plan specification. Interestingly, miRNA members belonging to families were expressed in overlapping tissues than miRNAs that do not belong to the same family. Together, the results provide evidence that at least a subset of intronic miRNAs may be controlled by their own, rather than a host gene promoter.

Essentially identical conclusions were also reached from studies of Drosophila miRNAs, whereby 15 "stand-alone" and 9 intronic miRNA loci and clusters were detected, collectively representing 38 miRNA genes.[39] A particular intronic miRNA, miR-7, showed a spatial expression pattern distinct from that of its host mRNA (bancal), suggesting an independent cis-regulatory control and hence an independent promoter. Recently, a more detailed study was conducted of the miR-281 gene of Drosophila, which revealed that it is located in the first intron of isoform RA of the ornithine decarboxylase antizyme (ODA) gene.[40] ODA has three isoforms because of alternative transcription start sites, but expression profile analysis indicated that miR-281 is not co-expressed with any of them. Indeed, transcription of pri-miR-281 initiated from its own promoter using hypophosphorylated RNA polymerase II (Pol II) and the transcription factor Myc, and terminated at a canonical polyadenylation site.

Thus, Class I intronic miRNAs represent a gene-within-a-gene. Regardless of their functional interaction with the host gene product, they may have regulatory relationships. A significant portion of the eukaryotic chromatin is transcriptionally activated in broad clusters such that multiple neighboring promoters are turned on simultaneously.[41] This can promote simultaneous activation of the Class I intronic miRNA gene and its host gene. Actual transcription will then be determined by the availability and activation of the cognate transcription factors. If the transcription factor(s) is (are) also shared by the host and the miRNA promoters, the same environmental signal can lead to joint transcription of both.

However, in such an event, the transcription elongation complex originated at the upstream host gene promoter (P_H in Fig. 1) and traveling through the downstream miRNA promoter (P_M in Fig. 1) may actually inhibit transcription initiation at the latter, a phenomenon termed "promoter occlusion".[42] The occlusion of the miRNA promoter will be more pronounced if the host promoter is considerably stronger.

The degree of complexity of the intronic miRNA promoters is currently unknown. High complexity with multiple enhancer sites is likely if these miRNAs are regulated by multiple environmental and cellular cues. The fact that there is no miRNA that is equally expressed in all cell types indeed suggests that their promoters are at least as stringently regulated as the protein-coding genes.[7]

2.2. Class II intronic miRNA

An arguably more interesting class, these intronic miRNA genes lack their own promoters and are transcribed as part of the host gene transcript. Thus, this class of intronic miRNAs are only generated by processing of the host mRNA precursor, and thus, constitute true examples of a regulatory non-coding RNA within a coding RNA. In theory, these miRNAs can be liberated from the precursor RNA by traditional RNA splicing. However, available evidence points to a novel mechanism in which they are processed from unspliced RNA intronic regions before splicing catalysis, likely between the splicing commitment step and the excision step.[4]

The Class II intronic miRNAs have been appreciated only recently and there are only a few reports describing their roles. Nonetheless, their joint biogenesis with their host mRNA offers intriguing new avenues of co-regulation that are summarized here.

First of all, due to their co-expression with host mRNA, the two have identical tissue-specific and cell-specific distribution. In the first example,[8] miR-126-5p, produced from the ninth intron of the EGF-like domain 7 (Egfl7) gene, is only found in tissues that also express Egfl7. Also known as VE-statin (vascular endothelial statin) or Neu1, Egfl7 is the first identified inhibitor of mural cell (vascular smooth muscle cells and pericytes) migration.

Surprisingly, selective chromosomal deletions of Egfl7 and miR-126 alleles revealed that the Egfl7 (-/-) mice were phenotypically normal, but the miR-126 (-/-) mice exhibited either embryonic death or postnatal retinal vascular phenotype.[43] Regulation of angiogenesis by miR-126 was confirmed by endothelial-specific deletion and in the adult cornea micropocket assay. These results document that an endothelial Class II intronic miRNA assist the function of its host gene product, although the mechanism remains to be unraveled. This study also cautions that inadvertent miRNA loss can be a complication of animal knockout strategies.

As expected, non-endothelial tissues such as the prostate are naturally deficient in Egfl7 gene expression, and as a result, contain very little miR-126.[8] In contrast, the prostate cells are famously enriched in a number of "prostate-specific antigens" that includes a novel protein, called prostein (also known as prostate cancer-associated protein 6 (PCANAP6) and solute carrier family 45, member 3 (SLC45A3).[44] It was shown that the absence of miR126 in the prostate is an essential pre-requisite for the ability of prostate cells to synthesize prostein.[8] Surprisingly, although prostein is expressed in both normal and cancerous prostate, ectopic expression of miR126 led to a significant reduction in the migration of the prostate cancer LNCaP cells *ex vivo*, suggesting that the natural lack of mir-126 can be a contributing factor to the invasiveness of prostate cancer in the appropriate genetic environment.[8]

We have recently proposed that a Class II intronic miRNA may assist the host gene function by silencing antagonistic mRNAs,[45] which makes prudent sense because the miRNA in this case is a byproduct of the host gene expression (Fig. 2). In the first recognized example,[45] it was found that transcriptional

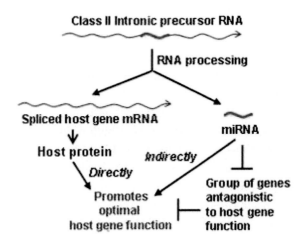

Fig. 2. Model for silencing of host-antagonistic genes by Class II intronic miRNA. Here, the miRNA produced from the common host gene transcript silences a group of genes that are antagonistic to the biochemical function of the host gene product.[45] Thus, the miRNA, which is a byproduct of the host gene, serves or facilitates the function of the host gene itself, creating a positive feedback. Note that this may also happen with some Class I intronic miRNAs, especially if the miRNA promoter (from P_M in Fig. 1) and the host gene promoter (from P_H in Fig. 1) are regulated by the same transcription factors.

activation of apoptosis-associated tyrosine kinase (AATK), essential for neuronal differentiation, also generated miR-338 from an AATK gene intron. Interestingly, miR-338 silenced a family of mRNAs whose protein products are negative regulators of neuronal differentiation, all of which had miR-338-specific MRE sequence(s) in their 3'-UTR.

The following were shown to be the major antagonistic targets of miR-338 and for good biochemical reasons[45]:

(i) UBE2Q1 that regulates remodeling and degradation of neuronal processes, including dendritic pruning, and is in fact implicated in Wallerian axonal degeneration, characteristic of conditions such as multiple sclerosis;

(ii) NOVA1, a regulator of brain-specific splicing, likely promotes the generation of splice variants of specific transcripts whose products suppress differentiation in the resting neuron;

(iii) DAB2IP, a tumor suppressor and general inhibitor of cell growth;

(iv) C2H2-171, a novel POZ-domain zinc finger protein (also known as ZNF238 or RP58, repressor protein 58 kDa), which is a transcriptional repressor in brain, and is also a developmental repressor.

As all these gene products are inhibitory to cell growth and proliferation in general, it makes biochemical sense for miR-338 to silence them in order to promote the agenda of its host AATK, which is to stimulate neuron growth. For the same reason, other potential targets of miR-338 that contain miR-338 MREs include the cerebellar degeneration-related protein 1 (CDR1) and the 26S proteasome non-ATPase regulatory subunit 7 (PSMD7).

Note that this regulation works in the form of RNA and is thus independent of the biochemical nature of the protein product of either the host gene or the antagonistic target genes.[45] For example, it would have worked just as well if AATK were say, a neural transcription factor rather than a kinase. The other notable feature of this mechanism is the simultaneous silencing of the full family of the target genes. In fact, the only requirement of this mechanism is that each target mRNA should have MRE sequence(s) for the same intronic miRNA in their 3'-UTR.

Location of the MRE in the 3'-UTR, as opposed to within the protein-coding sequence, allows greater flexibility of placement during the creation and evolution of the MREs. Indeed, many RNA regulatory elements are found in the 3'UTR that include the ARE that destabilizes RNA by recruiting RNases, the poly(A) stretch that binds poly binding protein (PABP), and the GAIT hairpin element that recruits a translation inhibitory complex containing ribosomal protein L13a to simultaneously silence a cohort of IFN-γ-induced pro-inflammatory gene mRNAs.[46,47]

Functionally, the GAIT complex is indeed highly reminiscent of the RISC complex recruited by the miR-MRE double-stranded sequence, revealing a common theme in translational regulation whereby cis-acting regulatory sequences in the noncoding region of the mRNA promote recruitment and assembly of multisubunit regulatory protein complexes. Recently, a family of jointly regulated mRNAs has been called a "post-transcriptional operon".[48] Clearly, the host-antagonistic mRNAs that are jointly regulated by a single Class II intronic miRNA and the proinflammatory mRNAs jointly regulated by the GAIT complex are notable examples of "post-transcriptional operon".

A survey of literature reveals possible other examples of such Class II intronic miRNAs functioning through this host-assisting mechanism.[45] For example, miR-151 resides within intron-21 of the gene for focal adhesion kinase (FAK), transcription of which is up-regulated in proliferative diseases such as cancer. It is thus prudent that a number of bioinformatically predicted targets of miR-151 are also regulators of cell cycle; in fact their expression is either expected or actually demonstrated to be reduced in various forms of cancer. Such targets include SCC-112, chromosome-associated kinesin KIF4A, and growth arrest-specific protein GAS-2.[45]

Another example is the homeobox (Hox) family genes, which are notable for their orderly activation during development. In zebrafish, genes Hox 1 through 6 are clustered in a single locus, such that when Hox4 protein appears, Hox1 and Hox3 proteins disappear. Interestingly Hox1 and Hox3 are both silenced by miR-10 that maps within an intron of Hox4.[49] Thus, miR-10, the intronic miRNA from the Hox4 gene, acts as a helper of its own host gene.

The very mechanism of biogenesis of the Class II intronic miRNAs ensures that they will be available to assist the host gene whenever the latter is activated, but at no other time.[45] However, the concentration of the intronic miRNA may discriminate between high- and low-affinity MREs in the antagonistic mRNAs and thereby fine-tune the silencing of each. Regardless, as the miRNA concentration will be regulated by the same signals and pathways that regulate the host gene promoter (e.g., RA, TPA, IGF for AATK), the expression of the miRNA is guaranteed to occur in proportion to that of the host gene, providing proportionate amount of help.

Reciprocally, when the host gene transcription will shut off upon cessation of the activating signal, the target protein levels will also be restored, thus providing a fast and reversible silencing mechanism. Also note that the model does not require that the host gene product will have a "positive" function and the target genes will be "negative"; in other words, the mechanism would work the same way if it were the reverse, i.e., if the host gene were a repressor of a pathway and the targets were activators of the same pathway. Indeed, the antagonism of the target genes toward the host gene is the fundamental aspect of the mechanism.[45]

3. The Creation of an MRE: "The Kiss of the miRNA" Hypothesis

Currently there is no hypothesis regarding how the miRNAs and their cognate MREs were created together. In proposing a mechanism of directed creation, we would like to hypothesize a "miRNA-first, MRE-later" scenario in which miRNAs created their own MREs by a miRNA-primed event ("kiss of the miRNA"), later selected by functional fitness via standard Darwinian evolution.

The core of the hypothesis is that the large variety of repeat sequences in the eukaryotic chromosome gave rise to double-stranded RNA structures that were processed by primitive RNases to generate miRNA-like short RNA molecules. These molecules generated partially complimentary MREs by acting as primers in the initiation of DNA replication. In the modern cell, short RNA is also a primer for DNA polymerase, but later removed by endonucleolytic cleavage. In the primitive metazoan cell, they were perhaps not removed but reverse transcribed and recombined into the DNA template.

Alternatively, the miRNA could be first reverse transcribed and then integrated into DNA. As priming would involve the 3' end of the miRNA, this end must find various degrees of complementary to the template, and therefore, would replicate the same variability from gene to gene as an MRE. In contrast, the 5' end of the miRNA do not need to conform to the template sequence, and would be incorporated as a perfect DNA copy. This would serve to explain the perfect match of the MRE to the 5'-end of the miRNA, known as the "seed" sequence.

Finally, insertions that occurred in the noncoding regions of the genome, such as introns and intergenic regions, were more likely to survive as they did not disrupt the open reading frames and essential regulatory elements. Eventually, only those MREs that were transcribed and provided an overall biochemical benefit in being regulated by the cognate miRNA, were retained through evolution and remained fixed in the genome. Others were either deleted or mutated beyond recognition by the homology-based algorithms. Over time, multiple MREs in the same 3' UTR accumulated through the same mechanism and/or MRE duplication and led to the combinatorial complexity that is typical of miRNA-regulated genes.[50]

4. The Size of the Host-Antagonistic Gene Pool is Roughly 5

In understanding the role of Class II intronic miRNA, it is helpful to have even a rough estimate of the number of genes that could be antagonistic to its host gene function. This is central to the prioritization of the bioinformatically predicted miRNA targets, rationalization the biochemical role of the targets, assessment of the demands on the cellular RISC machinery and our understanding of miR-MRE interaction and co-evolution.

To obtain an approximate estimate of how many antagonistic genes a Class II miRNA must suppress in order to help its host gene function, we can take AATK as an example of the host gene that harbored miR-338 and silenced at least 5 antagonistic genes to promote optimal neurite growth.[45]

For simplicity, we make the following assumptions: (a) the host gene (AATK) itself is non-antagonistic; (b) all antagonistic genes equally antagonize the host gene function; (c) all antagonistic genes are targeted simultaneously by the miRNA (miR-338) and that the cellular RISC can be rate-limiting;[51-53] (d) the antagonistic genes are neutral to one other.

As first approximation, we used NetLogo, a decentralized, multi-agent programming and modeling environment. NetLogo was selected because of its simple language structure and its support for mobile, goal-based agents that can move over a defined grid. As illustrated (Fig. 3), a Red agent represents an antagonistic gene and a Blue agent represents each of the many pathways that are utilized by AATK to activate neurite growth.

All squares move in a random fashion to simulate the diffusible cytoplasmic movement of the proteins. When a Red collides with a Blue by chance, Blue is destroyed, simulating the antagonism. We started the simulation with a single Blue and variable number of Reds, and run it for fixed turns every time (3000, to obtain workable numbers).

During this period, the Blue agent replicates, such that the number of surviving Blue agents at the end of the run is a measure of neuronal growth. As seen in the graph (Fig. 3B), maximum neurite growth is equivalent to about 2300, produced in the absence of any antagonism (Red = 0). A single Red agent (antagonistic gene) reduced the length by about 22% (to 1800), and two Red agents reduced the length by 40% (to 1400). As we have shown, this is significantly below the optimal desired length of neurons. The results indicate that it is crucial to silence essentially all antagonistic genes to achieve full host gene function.

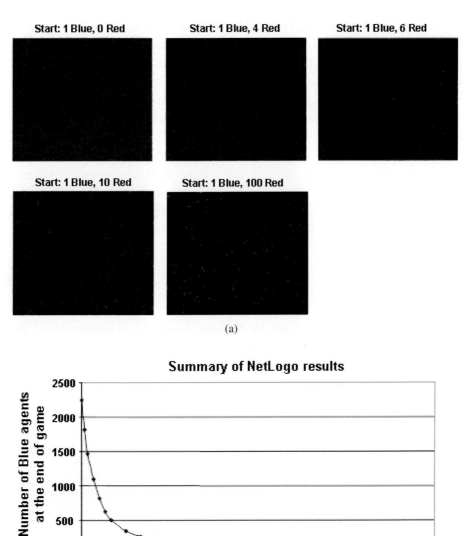

Fig. 3. NetLogo simulation to estimate the number of host-antagonistic genes. (a) Representative frame shots at the end of the simulation starting with the 1 Blue agent and 0-100 Red agents. Blue agents multiply and are eaten by Red agents, but the Red agents do not multiply (Details in Text). (b) Graphic plot of the total Blue agents (Y-axis) at the end of each simulation against the number of Red agents (X-axis). Each value was an average of three simulations; standard errors never exceeded ± 10% and were omitted for simplicity. Note that even 1 or 2 Red agents significantly inhibited the replication of Blue agents.

A cursory look at the graph (Fig. 3B) shows that the sharpest drop occurs till about 5-7 Red agents, following which the graph rapidly changes to a plateau. In other words, larger number of antagonistic genes is wasteful in terms of antagonistic effect. In the neuron growth study, we also achieved maximal antagonistic effect with 5 genes; reciprocally, fastest growth response was seen by antagonizing each of these genes using specific siRNAs, and full growth was achieved when all 5 were silenced by antagomir-338. Lastly, there is anecdotal evidence that silencing of multiple targets, in the order of 2 or 3, leads to lower RNAi response, most likely because the cellular RISC machinery is saturable. Thus, a given Class II intronic miRNA, when activated by host gene transcription, is unlikely to effectively silence more than roughly 5 antagonistic genes, which therefore defines the size of the antagonistic gene family for each host gene.

Acknowledgments

This work was supported in part by National Institutes of Health (USA) grant AI04583 (to SB).

References

1. Fire A, Xu S, Montgomery MK, Kostas SA, Driver SE, Mello CC. Potent and specific genetic interference by double-stranded RNA in Caenorhabditis elegans. Nature, 391, 806-11 (1998).
2. Ambros V. The functions of animal microRNAs. Nature, 431, 350-355 (2004).
3. Ruvkun G. The perfect storm of tiny RNAs. Nat. Med., 14, 1041-1045 (2008).
4. Kim YK, Kim VN. Processing of intronic microRNAs. EMBO J., 26, 775-783 (2007).
5. Lee EJ, Baek M, Gusev Y, Brackett DJ, Nuovo GJ, Schmittgen TD. Systematic evaluation of microRNA processing patterns in tissues, cell lines, and tumors. RNA, 14, 35-42 (2005).
6. Griffiths-Jones S, Saini HK, van Dongen S, Enright AJ. miRBase: tools for microRNA genomics. Nucleic Acids Res., 36(Database issue), D154-8 (2008), http://microrna.sanger.ac.uk/sequences/
7. Landgraf P, Rusu M, et al. A mammalian microRNA expression atlas based on small RNA library sequencing. Cell, 129, 1401-1414 (2007).
8. Musiyenko A, Bitko V, Barik S. Ectopic expression of miR-126*, an intronic product of the vascular endothelial EGF-like 7 gene, regulates prostein translation and invasiveness of prostate cancer LNCaP cells. J. Mol. Med., 86, 313-322 (2008).
9. Grimson A, Farh KK, Johnston WK, Garrett-Engele P, Lim LP, Bartel DP. MicroRNA targeting specificity in mammals: determinants beyond seed pairing. Mol. Cell, 27, 91-105 (2007).
10. Hutvagner G. Small RNA asymmetry in RNAi: function in RISC assembly and gene regulation. FEBS Lett., 579, 5850-5857 (2005).

11. Khvorova A, Reynolds A, Jayasena SD. Functional siRNAs and miRNAs exhibit strand bias. Cell, 115, 209-216 (2003).
12. Tomari Y, Matranga C, Haley B, Martinez N, Zamore PD. A protein sensor for siRNA asymmetry. Science, 306, 1377-1380 (2004).
13. Haley B, Zamore PD. Kinetic analysis of the RNAi enzyme complex. Nat. Struct. Mol. Biol., 11, 599-606 (2004).
14. Joshua-Tor L. The Argonautes. Cold Spring Harb. Symp. Quant. Biol. 71, 67-72 (2006).
15. Rivas FV, Tolia NH, Song JJ, Aragon JP, Liu J, Hannon GJ, Joshua-Tor L. Purified Argonaute2 and an siRNA form recombinant human RISC. Nat. Struct. Mol. Biol., 12, 340-349 (2005).
16. Liu J, Carmell MA, Rivas FV, Marsden CG, Thomson JM, Song JJ, Hammond SM, Joshua-Tor L, Hannon GJ. Argonaute2 is the catalytic engine of mammalian RNAi. Science, 305, 1437-1441 (2004).
17. Pillai RS, Artus CG, Filipowicz W. Tethering of human Ago proteins to mRNA mimics the miRNA-mediated repression of protein synthesis. RNA, 10, 1518-1525 (2004).
18. Pillai RS, Bhattacharyya SN, Filipowicz W. Repression of protein synthesis by miRNAs: how many mechanisms? Trends Cell Biol., 17, 118-126 (2007).
19. Filipowicz W, Bhattacharyya SN, Sonenberg N. Mechanisms of post-transcriptional regulation by microRNAs: are the answers in sight? Nat. Rev. Genet., 9, 102-114 (2008).
20. Meister G. miRNAs get an early start on translational silencing. Cell, 131, 25-28 (2007).
21. Förstemann K, Horwich MD, Wee L, Tomari Y, Zamore PD. Drosophila microRNAs are sorted into functionally distinct argonaute complexes after production by dicer-1. Cell, 130, 287-297 (2007).
22. Gusev Y. Computational methods for analysis of cellular functions and pathways collectively targeted by differentially expressed microRNA. Methods, 44, 61-72 (2008).
23. Zhang B, Wang Q, Pan X. MicroRNAs and their regulatory roles in animals and plants. J. Cell. Physiol. 210, 279-289 (2007).
24. Flynt AS, Lai EC. Biological principles of microRNA-mediated regulation: shared themes amid diversity. Nat. Rev. Genet., 9, 831-842 (2008).
25. Fabbri M, Croce CM, Calin GA. MicroRNAs. Cancer J. 14, 1-6 (2008).
26. Asli NS, Pitulescu ME, Kessel M. MicroRNAs in organogenesis and disease. Curr. Mol. Med. 8, 698-710 (2008).
27. Gaidatzis D, van Nimwegen E, Hausser J, Zavolan M. Inference of miRNA targets using evolutionary conservation and pathway analysis. BMC Bioinformatics. 8, 69 (2007) Erratum in: BMC Bioinformatics. 8, 248 (2007).
28. Baskerville S, Bartel DP. Microarray profiling of microRNAs reveals frequent coexpression with neighboring miRNAs and host genes. RNA 11, 241-247 (2005).
29. Lin SL, Miller JD, Ying SY. Intronic microRNA (miRNA). J. Biomed. Biotechnol., 2006, 26818 (2006).
30. Rodriguez A, Griffiths-Jones S, Ashurst JL, Bradley A. Identification of mammalian microRNA host genes and transcription units. Genome Res., 14, 1902-1910 (2004).
31. Weber MJ. New human and mouse microRNA genes found by homology search. FEBS J., 272, 59-73 (2005).
32. Ying SY, Lin S L. Intron-derived microRNAs – fine tuning of gene functions. Gene, 342, 25-28 (2004).

33. Thomson J, Newman M, Parker J, Morin-Kensicki E, Wright T, Hammond S. Extensive post-transcriptional regulation of microRNAs and its implications for cancer. Genes & Dev., 20, 2202-2207 (2007).
34. Borchert G, Lanier W, Davidson B. RNA polymerase III transcribes human microRNAs. Nat. Struct. Mol. Biol., 13, 1097-1101 (2006).
35. Cai X, Hagedorn C, Cullen B. Human microRNAs are processed from capped, polyadenylated transcripts that can also function as mRNAs. RNA 10, 1957-1966 (2004).
36. Lee Y, Kim M, Han J, Yeom K, Lee S, Baek S, Kim V. MicroRNA genes are transcribed by RNA polymerase II. EMBO J., 23, 4051-4060 (2004).
37. Ozsolak F, Poling LL, Wang Z, Liu H, Liu XS, Roeder RG, Zhang X, Song JS, Fisher DE. Chromatin structure analyses identify miRNA promoters. Genes Dev. 22, 3172-3183 (2008).
38. Martinez NJ, Ow MC, Reece-Hoyes JS, Barrasa MI, Ambros VR, Walhout AJ. Genome-scale spatiotemporal analysis of Caenorhabditis elegans microRNA promoter activity. Genome Res., 18, 2005-2015 (2008).
39. Aboobaker AA, Tomancak P, Patel N, Rubin GM, Lai EC. Drosophila microRNAs exhibit diverse spatial expression patterns during embryonic development. Proc. Natl. Acad. Sci. USA, 102, 18017-18022 (2005).
40. Xiong H, Qian J, He T, Li F. Independent transcription of miR-281 in the intron of ODA in Drosophila melanogaster. Biochem. Biophys. Res. Commun. 378, 883-889 (2009).
41. Tomancak P, Berman BP, Beaton A, Weiszmann R, Kwan E, Hartenstein V, Celniker SE, Rubin GM. Global analysis of patterns of gene expression during Drosophila embryogenesis. Genome Biol. 8, R145 (2007).
42. Adhya S, Gottesman M. Promoter occlusion: transcription through a promoter may inhibit its activity. Cell, 29, 939-944 (1982).
43. Kuhnert F, Mancuso MR, Hampton J, Stankunas K, Asano T, Chen CZ, Kuo CJ. Attribution of vascular phenotype of the murine Egfl7 locus to the microRNA miR-126. Development, 135, 3989-3993 (2008).
44. Xu J, Kalos M, Stolk JA, Zasloff EJ, Zhang X, Houghton RL, Filho AM, Nolasco M, Badaro R, Reed SG. Identification and characterization of prostein, a novel prostate-specific protein. Cancer Res., 61, 1563-1568 (2001).
45. Barik S. An intronic microRNA silences genes that are functionally antagonistic to its host gene. Nucleic Acids Res., 36, 5232-5241 (2008).
46. Vyas K, Chaudhuri S, Leaman DW, Komar AA, Musiyenko A, Barik S, Mazumder B. Genome-wide polysome profiling reveals an inflammation-responsive posttranscriptional operon in gamma interferon-activated monocytes. Mol. Cell. Biol., 29, 458-470 (2009).
47. Mazumder B, Seshadri V, Fox PL. Translational control by the 3'-UTR: the ends specify the means. Trends Biochem. Sci., 28, 91-98 (2003).
48. Keene JD. RNA regulons: coordination of post-transcriptional events. Nat. Rev. Genet., 8, 533-543 (2007).
49. Woltering JM, Durston AJ. MiR-10 represses HoxB1a and HoxB3a in zebrafish. PLoS ONE, 3, e1396 (2008).
50. Okamura K, Chung WJ, Lai EC. The long and short of inverted repeat genes in animals: microRNAs, mirtrons and hairpin RNAs. Cell Cycle, 7, 2840-2845 (2008).
51. Bitko V, Musiyenko A, Shulyayeva O, Barik S. Inhibition of respiratory viruses by nasally administered siRNA. Nat. Med., 11, 50-55 (2005).

52. Barik S. Control of nonsegmented negative-strand RNA virus replication by siRNA. Virus Res., 102, 27-35 (2004).
53. Hutvagner G, Simard MJ, Mello CC, Zamore PD. Sequence-specific inhibition of small RNA function. PLoS Biol. 2, E98 (2004).

CHAPTER 7

MIROME ARCHITECTURE AND GENOMIC INSTABILITY

Konrad Huppi[1], Natalia Volfovsky[2], Brady Wahlberg[1],
Robert M. Stephens[2] and Natasha J. Caplen[1]

[1]*Gene Silencing Section, Genetics Branch, Center for Cancer Research,
National Cancer Institute, National Institutes of Health,*
[2]*Advanced Biomedical Computing Center, Advanced Technology Program,
NCI-Frederick/SAIC Frederick Inc., NIH*

Genomic instability has been observed in many different types of cancers. While genetic alterations often cover a large spectrum of genomic events including gross rearrangement or amplification or deletion of chromosomal regions, some alterations are minimally detectable including small microsatellite or indel mutations and retroviral integration. In fact, the study of events associated with genomic instability has led to the discovery of many new and important genes associated with critical regulatory pathways. More recently, small regulatory RNAs or microRNAs have been found to reside throughout the genome and it has been suggested that such genetic alterations could also target microRNAs. In this chapter we will examine genomic instability that may alter human or mouse microRNA expression and function. Using both computational and experimental approaches, we outline possible associations between regions of genomic instability and novel microRNA candidates or already established and known microRNAs.

1. Introduction

Genetic instability in mammals has primarily been associated with fragile sites or regions with high rates of transcription that tend to be hotspots for translocations, integration of exogenous nucleic acids and gene amplification or deletion.[1] Perhaps the single most common denominator in genomic instability is the generation of DNA breaks. Thus, the inability of certain proteins to prevent or suppress the proper repair of these DNA breaks is yet another significant contributing factor in the generation of genetic instability. The observation that many of these breakpoints are associated with cancer related or developmentally important genes has become an important gene discovery tool for molecular biologists.[2]

For example, the alignment of immunoglobulin genes with MYC t(8:14) translocations in Burkitt's lymphoma, BCL2 t(14:18) translocations in Follicular lymphoma or the classic Philadelphia chromosome t(9:22) translocation that misaligns ABL with BCR in chronic myelogenous leukemia have lead to many important contributions to the cancer biology field as well as the discovery of the proto-oncogenes MYC, BCL2 and ABL.[3] In fact, finding genomic counterparts to the transforming viruses v-ras, v-abl, v-myc and v-src in the mammalian genome, led to the oncogene hypothesis that mutations in virally transmitted genes or oncogenes contribute to the transforming phenotype and malignancy.[4]

Among integrating viruses, the human papilloma virus (HPV) is found to target fragile sites, whereas others such as human immunodeficiency virus (HIV) appear to harness the cellular properties associated with actively transcribing regions.[5] This apparent preference for integration sites would support the concept that viral integration is not random, and can often follow patterns similar to that observed in chromosomal translocations and amplification/deletion wherein the breaks are in proximity to transcribed genes (i.e. directed gene targeting).

Consistent with this observation, studies of mouse tumors generated by retroviral integration of MLV have lead to the discovery of many targeted genes and remains an important cancer associated gene discovery tool.[6,7] Similarly, amplification and deletion of chromosomal regions detected largely through comparative genome hybridization (CGH) has become yet another means of identifying genes on a whole genome basis which may be involved in the cancer process. Particularly in the case of amplification/deletion and in some cases of viral integration or chromosomal translocation, the apparent target gene is not immediately obvious nor does the target gene appear to have properties considered to be indicative of a cancer gene target (i.e. β-Actin).

However, the recent discovery of small non-coding RNAs or miRNAs now may suggest reasonable alternative targets. Ironically, the miRNAs may have been originally dismissed as insignificant targets due to their small size, but increasing evidence suggests their expression levels may have an even greater impact due to the fact that they can target hundreds of genes. Furthermore, miRNAs are actively transcribed (by Pol II), so they could conceivably be exposed for long periods of time providing ideal targets for viruses or chromosomal translocation. Even in the presence of nearby protein coding genes, it is possible that miRNAs may in fact, be the intended target of viral integration or chromosomal translocation events. In this chapter, we review aspects of the miRome architecture particularly in reference to cancer phenotypes associated with genomically unstable human and experimental mouse regions.

2. Genomic Landscape of the miRome

Sequences corresponding to miRNAs are found interspersed throughout the genomes of virtually all organisms including plants, viruses and mammals. The genes encoding miRNAs are dispersed throughout every chromosome except Y, and it has been noted that a significant proportion (68.5% in the mouse) are found in clusters as defined by physical association, i.e. within 1 Mb, and/or by biological association (i.e. under the same regulatory control) (KH manuscript in prep.).

With rare exceptions, miRNAs are also found to be largely conserved across synteny, further strengthening the observation that they are distributed non-randomly. In the mouse, there are 75 clusters of miRNAs including 38 that are paired such that 68.5% of all miRNAs can be found within a cluster. Several miRNA clusters in the mouse contain five or more miRNAs and these clusters are located on the proximal end of Chr. 2 (9 miRNAs), Chr. 3 (5 miRNAs), the distal end of Chr. 5 (5 miRNAs), the proximal end of Chr. 6 (5 miRNAs), Chr. 8 (5 miRNAs), Chr. 12 (47 miRNAs), Chr. 14 (6 miRNAs) and Chr. 15 (7 miRNAs). On Chr. X there are three large clusters of miRNAs of 5, 13 and 6 members.[8]

Similar clusters are also found in humans and on the same syntenic chromosomal regions.[9] Within some clusters of miRNAs, members of the same miRNA family can be found such as miR-17 and miR-20a, both present within the miR-17~92 cluster on mouse chromosome 14. The fact that the miR-17~92 cluster also contains miRNA members of other families as well argues against simple gene duplication as the mechanism for generating this cluster. However, the largest cluster of miRNAs (on mouse chromosome 12 and human chromosome 14) contains many members of the same miRNA family recently duplicated in such a fashion as to make it difficult to distinguish miRNAs.

Primary miRNAs are transcribed by RNA polymerase II (Pol II) using either their own promoters or those associated with neighboring protein-encoding genes. The long primary transcripts of miRNAs are processed into 65-80 nucleotide (nt) hairpin structures (precursor miRNAs) by the ribonuclease complex formed by DROSHA and DGCR8 and it is believed that many clustered miRNAs may actually be transcribed as one primary transcript which is then spliced into individual miRNA specific precursor transcripts.

It is estimated that ultimately, several hundred to a thousand miRNAs might exist in a particular mammalian organism. Currently, 695 human and 488 mouse miRNAs have been assigned (miR-Base release 12.0, September 2008). A detailed analysis of the location of miRNAs relative to the start and end positions of annotated transcriptional units within the mouse genome shows that 93% of all miRNAs are either intragenic (44%) or intergenic and very close (within 100kb) to a larger transcript (49%) (KH, manuscript in prep.).

Finding a significant majority of miRNAs located within transcriptional boundaries of protein coding genes is another strong indicator that many miRNAs are controlled by the transcriptional regulators associated with these larger gene units. For example, miR-155, which resides within the BIC1 locus, appears to be the target of MLV induced lymphoma in chickens.[10] Further expression studies in lymphocytes show coordinately regulated expression of both BIC1 and miR-155.[11,12]

In another example, expression of both miR-338 and the surrounding Aatk gene on mouse chromosome 11 have been shown by northern blotting to be coordinately regulated.[13] Although the direct confirmation of co-regulated expression of miRNAs and the larger transcripts that surround them is limited, it is anticipated that this observation will become commonplace as miRNAs are further studied. Similarly, it is expected that the expression of clustered miRNAs will often be found in a co-regulated fashion. For example, the miR-1-1/miR-133 cluster of miRNAs found on mouse chromosomes 2 and 18 have been shown to be coordinately expressed in skeletal muscle and heart.[14]

Thus, we view transcriptional regulation as an intricate series of interactions between promoters, enhancers, insulators and other episomal elements and we can expect that any interruption of the larger gene units, through chromosomal translocation or retroviral integration will clearly affect the regulation of nearby miRNAs as well.

3. miRNAs and Chromosomal Translocations

Years ago, it was noted that an aggressive B cell leukemia with a t(8;17) chromosomal translocation appeared to activate MYC in association with an unknown gene located on chromosome 17.[15] At the time, no recognizable gene transcript was found in the immediate region of the breakpoint. More recently, the discovery of miRNAs led to a more intensive search of this region and the realization that the breakpoint resided only 50nt from the 3' end of the mature miR-142 sequence and 4nt from the 3' end of the pre-miR-142.[16] Thus, it is thought that the proximity of the breakpoint possibly interrupts the regulation of mir-142.

In another example, a meningioma with a t(4;22) translocation has a chromosomal breakpoint only 1kb from miR-180 which appears to activate both the involved MN1 gene and miR-180.[17] In a patient with B cell acute lymphoblastic leukemia (B-ALL), insertion of miR-125b-1 into the immunoglobulin heavy chain (IgH) locus results in the activation of miR-125b-1 under control of the IgH enhancer.[18] A region downstream of MYC in both humans and mouse is susceptible to breakage in Burkitt's lymphoma t(8:22) or t(2:8) translocation or mouse plasmacytoma t(6:15) or t(15:16) translocation, in which one of the immunoglobulin light chain loci is brought into proximity with a region downstream of MYC.[19]

Although transcripts encompassing this region (PVT1) were isolated in both human and mouse years ago, the fact that they were found to be noncoding encouraged our laboratory to search for new and as yet undetected miRNAs that might explain the targeting of translocation breakpoints.[20] A 400kb sequence defining most of the PVT1 region (University of California, Santa Cruz (UCSC) hg17, chr8: 128,830,208-129,253,652) was examined in detail for predicted stable hairpin formation, and sequence conservation among species. Conserved segments of 60bp or more were extracted from the UCSC PhastCons based on alignment from six species including human chimpanzee, dog, mouse, rat and chicken (with a sliding window of 80bp in increments of 10bp). Sequences were then subjected to folding analysis using RNAfold. Selected structures with a length greater than 50 bp and a minimum of 65% base pairing were subjected to further folding analysis and only those with a free energy over –15 kcal/mol were considered candidate miRNAs. By this criteria, seventeen candidates were identified which where subjected to further analysis by northern hybridization to small purified RNAs from several representative cell lines. Several miRNAs (miR-1204~1208) were positively identified with these parameters and further analyzed by 5' RACE and primer extension.

From these studies, primers could be designed for RT-PCR quantitation of precursor and mature transcripts. These miRNAs appear to be coordinately regulated among themselves and with the surrounding PVT1 transcript in developing B cells as well as in B cell tumors. Furthermore, studies of MLV induced T cell lymphomas reveals one of the most frequent RIS is the PVT1 locus.[21] Increased levels of expression of both PVT1 and the associated miRNAs are observed in most of the T cell tumors.

4. miRNAs and Chromosomal Amplification or Deletion

The human 8q24 region is not only susceptible to chromosomal translocation but is also a part of gross chromosomal amplification in a large number of malignancies including breast, ovarian, lung, head and neck, prostate, colon, myeloma and neuroblastoma.[22]

We have also found consistently increased levels of expression of miR-1204~1208 in several tumors carrying amplified segments of the 8q24 region including breast and colon.[20]

Another chromosomal region amplified in a number of lymphomas including Diffuse Large B cell lymphoma (DLBL), T cell acute lymphoblastic lymphoma (T-ALL) and mantle cell lymphoma to name a few is chromosome 13q31 which harbors the miR-17~92 cluster.[23] Northern analysis shows that 6/7 of the miRNAs (miR-18 being the exception) from the cluster are over-expressed in samples with a gain of 13q31.

Transduction of the portions of miR-17~92 cluster into deficient or under-expressed cell lines with over-expressing plasmid or viral constructs results in lymphoproliferative disease such as found in the B or T cell lymphomas.[24,25] However, the fact that only certain parts, not the entire segment of the miR-17~92 cluster is necessary to produce this phenotype suggests that there may be some redundancy among miRNAs in this cluster.

Also important is the fact that eliminating the miR-17~92 locus in mice by genetic knockout results in not only a lymphocyte developmental blockage at the pro to pre-B cell stage as expected, but also a septal heart defect that render the mice embryonically lethal.[26] This duality in roles emphasizes the many potential interactions of miRNA clusters, particularly those with several members.

In a comprehensive pediatric T-ALL study of chromosomal gain/loss, the entire chromosome 8 is amplified including the miR-1204~1208 cluster and among regions of loss, there are a number of miRNAs including miR-198 (3q13), miR-569 (3q26), 11 miRNAs (7q31), miR-101-2 (9p24), 10 miRNAs (9q21), miR-604 (10p11), miR-613/614 (12p13), miR-15/16a (13q14), miR-193a/365-2 (17q11), miR-199a (19p13) and miR-103-2 (20p12) [27]. The region of 7q32 is also deleted in many cases of splenic marginal zone lymphoma with specific reduction in expression miR-29a/29b-1 cluster.[28]

One recognized target of miR-101 is the enhancer of zeste homolog 2 (EZH2), which is found to be elevated in a number of solid tumors. In a recent study of prostate tumors,[29] lower levels of expression in miR-101 appears to be associated with genomic loss of both loci of miR-101 on human chromosome 1 (miR-101-1, 12-51%) and chromosome 9 (miR-101-2, 24-25%). Several phenotypic assays involving inhibition or over-expression suggest a role for miR-101 in the progression to more aggressive forms of disease including metastasis.

As one of the burgeoning areas of research, we anticipate that correlating higher resolution whole genome analysis using CGH in cancer to transcription of large transcripts and miRNAs will significantly improve our understanding of chromosomal gain/loss. For example, whole genome scanning of human lung cancer lines with detailed analysis (~100kb resolution) of chromosome deletion reveals downregulation of protein coding genes as well as three miRNAs, miR-99a, let-7c, and miR-125b-2 within the deleted 21q21 region.[30,31] Although limited to a single patient with a rare embryonal tumor with abundant neutrophil and true rosettes (ETANTR), a recent study using array CGH identified several regions of gain/loss across a number of chromosomes including genomic amplification of a 0.89Mb region on chromosome 19q13. Using genome-wide miRNA profiling, the authors showed that among the highest expressing miRNAs were the cluster of miRNAs (miR-518b, miR-519d, miR-520-3p, miR-517, miR-372 and miR-373) that reside within the amplified segment.[32]

5. Retroviral Integration and the Mouse miRome

A few years ago, we and others noted that there is a degree of co-localization in the mouse genome between the position of miRNAs and the sites of retroviral insertion (RIS) seen in mouse tumors.[8,33] The assignment of a RIS clone to a marker locus is dependent on the available gene/transcript based annotation. We were interested to see to what degree the recent discovery of miRNAs may impact the assignment of gene markers to RIS clones and thus the potential underlying biology of these mutagenic insertions.

The latest update of the RTCGD has four RIS assigned to miRNA genes including two insertional clones from a NUP98-HOXD13 mouse model that map close to miR-29a on Chr. 6,[34] and two Dkmi18 clones associated with B-cell tumors that map within 20 kb of miR-let7g on Chr. 9. A set of retroviral insertion clones seen within a specific region on mouse Chr.14 is quite remarkable in that the RIS clones found in this region were originally associated with the GPC5 gene as this was the closest annotated gene at that time.

More recent studies, including the analysis of a series of SL3-3 induced T-cell tumors,[35] shows that these integration sites are actually closer to a cluster of four miRNAs (miR-17, -18, -19a and -20a) upstream of GPC5. In fact, one RIS (Lnz25-3) resides only 945bp proximal to miR-17 on Chr. 14.

In another example, a RIS (HH76) is located only 1.6kb from the miR-92-2/106a cluster of miRNAs on mouse Chr. X with no known protein coding gene nearby.[36] In a mouse model of hepatocellular carcinoma, integration of an adeno-associated virus (AAV) was found to be within a 6kb region with two of the RIS just 12bp apart and within the miR-341 transcript which is part of the large cluster of miRNAs found on mouse chromosome 12.[37]

We have compiled in Tables 1 and 2 all the current mouse and human miRNAs that map within 100kb of retroviral integration sites along with the chromosome location and closest gene, distance from the integration site, tumor type and the retrovirus construct. From this table we have found that the viral construct is critical to what type of tumor is generated, so there may be a bias in what miRNAs may be targeted.

However, it is quite clear that several miRNA clusters are frequently involved in the RIS event including miR-1204~1208 on mouse chromosome 15, miR-106a~363 on chromosomes X, miR-142 on chromosome 11, miR-200c, miR-29a/29-b-1 and miR-196 all on chromosome 11. In some cases, the miRNAs are found extremely close to the integration sites. Some examples are miR-9-1 which is only 4.4kb from a RIS on chromosome 3 that generates a brain CA and miR-200c which is only 12.3kb from the RIS on chromosome 6 which generates B cell lymphomas.

The Hox locus on mouse chromosome 6 also harbors a number of RIS that are found interspersed between Hox A7, A9 and A10, but the fact that miR-196 is only 296bp from one RIS with several other RIS very close by would argue that the targeting very likely is miRNA as opposed to targeting of the protein coding Hox genes. On mouse chromosome 14, there are no protein coding genes close to the RIS which are found as close to miR-17~20a as 298bp and 771bp in erythroleukemia or T lymphoma, respectively. Overexpression of the entire miRNA cluster is further evidence that the miRNAs appear to be solely responsible for the tumors associated with these integration events.[35]

As mentioned previously, Wabl and colleagues were able to induce T cell lymphomas which carry a number of MLV RIS into the region transcribing PVT1 and the miR-1204~1208 cluster on mouse chromosome 15.[21] As expected, higher levels of expression of both PVT1 and the closer miRNAs is observed[21] However, the chromosome 15 RIS were often accompanied by RIS in other chromosomal regions including protein coding genes (i.e., Evi5, Notch1, Rasgrp1) and miRNAs (miR-106a~363).[36,37] In an example on mouse chromosome 13, the miR24-1~23b cluster is flanked by two RIS that generate brain CA. The RIS are as close as 6.3kb to the miRNAs and yet, they are also close to the coding gene Fancc. However, finding the RIS located downstream of the Fancc promoter region in this case would indicate that the RIS event probably targeted the miRNAs as promoters are typically unidirectional.

Table 1. Mouse microRNAs and chromosomal breakpoints.

mmu-miRNA	CHR location	Closest distance	Closest gene	Chromosomal break	Virus	Tumor	Refs.
miR-375	1	84.0kb	*Wnt6*	retrovirus (4)	AKV	B cell lymphoma	38
miR-350	1	42.1kb	*Sdccag8*	retrovirus (2)	AKV	B cell lymphoma	38
miR-29c	1	93.1kb	*Crry*	retrovirus (2)	AKV	B cell lymphoma	38
miR-669b	2	95.9kb	*Sfmbt2*	retrovirus (2)	MMTV	mammary tumor	39
miR-126	2	41.4kb	*Notch1*	retrovirus (33)	MLV	T, B lymphoma	40
miR-130a	2	99.2kb	*Slc43a1*	retrovirus (3)	AKV	B cell lymphoma	38
miR-9-1	3	4.4kb	*Rhbg*	retrovirus (3)	PDGF-MLV	Brain Ca	41
miR-101a	4	48.8kb	*Jak1*	retrovirus(4)	AKV	B cell myeloid	42, 38
miR-339	5	69.2kb	*Cox19*	retrovirus (3)	AKV	B cell lymphoma	43
miR-29a/29b-1	6	18.6kb	*Hoxa7*	retrovirus(12)	MLV	B cell lymphoma	44
					PDGF-MLV	Brain	44
				retrovirus	MLV	leukemia	45
				retrovirus	AKV	lymphoma	38
				retrovirus	MOL4070	stem	34
miR-196b	6	290bp	*Hoxa9*	retrovirus (25)	MLV	B cell	46
					AKV	T cell	43
					AKV	myeloid	47
miR-200c	6	12.3kb	*Ptpn6*	retrovirus (4)	AKV	B cell	38
				retrovirus (2)	MLV	HS	45
miR-9-3	7	71.8kb	*Rhcg*	retrovirus (3)	PDGF-MLV	Brain	41
				retrovirus (1)	MLV	leukemia	40

Table 1 (*Continued*)

mmu-miRNA	CHR location	Closest distance	Closest gene	Chromosomal break	Virus	Tumor	Refs
miR-483	7	11.5kb	*Igf2*	retrovirus (3)	MMTV	mammary tumor	39
miR-135a-1/let-7g	9	6.0kb		retrovirus (2)	MLV	B lymphoma	44
miR-132/212	11	57bp	*Hic1*	retrovirus (3)	AKV	B lymphoma	38
				retrovirus (2)	MLV	T lymphoma	48
miR-142	11	within		retrovirus (8)	MLV	T, B lymphoma	40, 44, 48
				retrovirus (1)	AKV	myeloid	43
				retrovirus (1)	MLV	HS	45
miR-341/370	12	within	*Gtl2*	adenovirus (4)	AAV	HCC	37
miR-24-1/27b/23b	13	6.3kb	*Fancc*	retrovirus (2)	PDGF-MLV	Brain	41
miR-17/18/19a/20a	14	771bp		retrovirus (7)	SL3-3	T lymphoma	35
		298bp		retrovirus (?)	F-MLV	erythro-leukemia	50
				retrovirus (1)	AKV	B cell	43
miR-1204/1205/1206/1207/1208	15	6bp	*Pvt1*	retrovirus (22)	MLV	T lymphoma	21
		1kb		T(6;15) (20)		plasmacytoma	49
miR-219-1	17	96.2kb	*Brd2*	retrovirus (1)	AKV	B cell	43
				retrovirus (1)	MLV	B cell	44
miR-194-2	19	75.5kb	*Map4k2*	retrovirus (3)	AKV	B cell	43
miR-146b	19	33.2kb	*NFkb2*	retrovirus (3)	AKV	B cell	43
miR-106a/20b/19b-2/92-2/363	X	1761bp		retrovirus (13)	MLV	B T lymphoma	43
				retrovirus (23)	SL3-3	T lymphoma	36

Table 2. Human microRNAs and chromosomal breakpoints.

hsa-miRNA	CHR Location	Closest Distance	Closest Gene	Chromosomal break	Tumor	Refs.
miR-28	3			3q27 translocations	CNS lymphoma	51
miR-1204/ 1205/1206/ 1207/1208	8	10bp	PVT1	T(8;22), T(2;8)	Burkitt Lymphoma	20
let-7i	12			12q15 translocations	Benign CA	52
miR-15/16-1	13	within	DLEU2	mutation	B-CLL	16
				13q14 deletion	B-CLL	16
				T(2;13)	CLL	16
miR-125b-1	14	within	IgH	14q32	B-ALL	17
miR-142	17	within	SupT4H	T(8;17)	B- lymphoma	16
		within		17q23 amplicon	Breast CA	17
		? kb		HPV-16	Cervical CA	17

Of particular interest are a series of RIS associated with B cell or T cell lymphomas that are found clustered between two coding region genes, Hic1 and Ovca1 on mouse chromosome 11. Complete disruption of the Hic1 allele renders the mice embryonically lethal while knockout of a single allele leads to viable mice that were susceptible to certain types of cancer depending on the sex of the mouse.[53]

This study showed that hypermethylation of one of two alternative promoters of Hic1 leads to silencing of that promoter and a corresponding gender-dependent spectrum of malignant tumors. Female mice tended towards lymphomas and sarcomas whereas male mice developed epithelial cancers. Studies to determine why Hic1 promoter specific hypermethylation leads to gender- and cancer-specific differences are ongoing. Our own studies reveal two miRNAs (miR-212, miR-132) are located between Hic1 and Ovca1 and in close proximity to the alternative promoters used for Hic1. Thus, we would suggest that miR-212/132 could represent an alternative target of the silencing effect of hypermethylation that could lead to the wide range of effects observed in the original studies of Baylin's group. The fact that six RIS breakpoints associated with T cell or B cell lymphomas are found extremely close (57-2000bp) and appear to target miR-212/132 as opposed to Hic1 or Ovca1, further substantiates the hypothesis that these miRNAs may be the intended targets of any genomic changes to this region.

6. Conclusions and Outlook

If estimates are essentially correct that ~1000 miRNAs may be expected within the mammalian organism, then the current miRNA count lies at roughly 50% of the expected total. Whole genome sequencing combined with dense tiling expression arrays may prove to be the surest way of identifying most of the remaining miRNAs. There may also be a bias in the current library towards non-telomeric or centromeric based miRNAs until contiguous sequences are obtained from these regions.

In the meantime, focused studies on genomically unstable regions, particularly those that are devoid of large coding region transcripts represent a rich resource for linking known miRNAs or identifying previously unknown miRNAs.

Acknowledgements

We wish to acknowledge support of the intramural research program (Center for Cancer Research, NCI) of the NIH and Federal funds from the National Cancer Institute, National Institute of Health, under Contract HHSN261200800001E.

References

1. A. Aguilera and B. Gomez-Gonzalez, Genome instability: a mechanistic view of its causes and consequences, Nature Rev., 9, 204-217, 2008.
2. J. C. Neil and E.R. Cameron, Retroviral insertion sites and cancer: Fountain of all knowledge? Cancer Cell, 2, 253-255, 2002.
3. C. E. Gauwerky and C.M. Croce in Molecular Basis of Cancer, J. Mendelsohn, P.M. Howley, M.A. Israel, L.A. Liotta eds. pp. 18-37, 1995.
4. J.M. Bishop, The molecular genetics of cancer, Science, 235, 305-311, 1987.
5. F. Bushman, M. Lewinski, A. Ciuffi, S. Barr, J. Leipzig, S. Hannenhali and C. Hoffmann, Genome-wide analysis of retroviral DNA integration, Nature Rev., 3, 848-858, 2005.
6. L.S. Collier and D.A. Largaespada, Transforming science: cancer gene identification, Cur. Opinion in Genetics & Dev., 16, 23-29, 2006.
7. I.P. Touw and S.J. Erkeland, Retroviral insertion mutagenesis in mice as a comparative oncogenomics tool to identify disease genes in human leukemia, Mol. Therapy, 15, 13-19, 2007.
8. K. Huppi, N. Volfovsky, M. Mackiewicz, T. Runfola, T.L. Jones, S.E. Martin, R. Stephens and N.J. Caplen, MicroRNAs and genomic instability, Sem. in Cancer Biol. 17, 65-73, 2007.

9. G.A. Calin, C. Sevignani, T. Dumitru, E. Hyslop, S. Noch, S. Yendamuri, M. Shimizu, S. Rattan, F. Bullrich, M. Negrini and C.M. Croce, Human microRNA genes are frequently located at fragile sites and genomic regions involved in cancers, Proc. Natl. Acad. Sci. USA, 101, 2999-3004, 2004.
10. W. Tam, D. Ben-Yehuda and W.S. Hayward, bic, a novel gene activated by proviral insertions in avian leukosis virus-induced lymphomas, is likely to function through its noncoding RNA, Mol. Cell. Biol. 17, 1490-1502, 1997.
11. J. Kluiver, S. Poppema, D. De Jong, T. Blokzijl, G. Harms, S. Jacobs, B-J. Kroesen and A. van den Berg, BIC and m iR-155 are highly expressed in Hodgkin, primary mediastinal and diffuse large B cell lymphomas, J. of Pathology 207,243-249, 2005.
12. M. Metzler, M. Wilda, K. Busch, S. Viehmann, A. Borkhardt, High expression of precursor microRNA-155/BIC RNA in children with Burkitt lymphoma, Genes, Chromosomes & Cancer, 39, 167-169, 2004.
13. S. Barik, An intronic microRNA silences genes that are functionally antagonistic to its host gene, Nucleic Acids Res., 36, 5232-5241, 2008.
14. J-F. Chen, E. M. Mandel, J.M. Thomson, Q. Wu, T.E. Callis, S.M. Hammond, F.I. Conlon and D-Z. Wang, The role of microRNA-1 and microRNA-133 in skeletal muscle proliferation and differentiation, Nat. Genet., 38, 228-233, 2006.
15. C.E. Gauwerky, K. Huebner, M. Isobe, P.C. Nowell and C. M. Croce, Activation of MYC in a masked t(8;17) translocation results in a aggressive B-cell leukemia, Proc. Natl. Acad. Sci. USA, 86, 8867-8871, 1989.
16. G. A. Calin , C. Sevignani, C.D. Dumitru, T. Hyslop, E. Noch, S. Yendamuri, M. Shimizu, S. Rattan, F. Bullrich, M. Negrini and C.M. Croce, Human microRNA genes are frequently located at fragile sites and genomic regions involved in cancers, Proc. Natl. Acad. Sci. USA, 101, 15524- (2002).
17. G. A. Calin and C. M. Croce, Chromosomal rearrangements and microRNAs: a new link with clinical implications, J. Clin. Invest. 117, 2059-2066, 2007.
18. T. Sonoki, E. Iwana, H. Mitsuya and N. Asou, Insertion of microRNA-125b-1, a human homologue of lin-4, into a rearranged immunoglobulin heavy chain gene locus in a patient with precursor B-cell acute lymphoblastic leukemia, Leukemia, 19, 2009-2010, 2005.
19. M. Potter, Neoplastic development in plasma cells, Immunol. Rev. 194, 177-195, 2003.
20. K. Huppi, N. Volfovsky, T. Runfola, T.L. Jones, M. Mackiewicz, S.E. Martin, J. F. Mushinski, R. Stephens and N. J. Caplen, The identification of microRNAs in a genomically unstable region of human chromosome 8q24, Mol. Cancer Res. 6, 212-221, 2008.
21. G. B. Beck-Engeser, A. M. Lum, K. Huppi, N.J. Caplen, B.B. Wang and M. Wabl, Pvt1-encoded microRNAs in oncogenesis, Retrovirology, 5, 4, 2008.
22. Y. Guan, W.L. Kuo, J.L. Stilwell, H. Takano, A.V. Lapuk, et al., Amplification of PVT1 contributes to the pathophysiology of ovarian and breast cancer, Clin. Cancer Res., 13, 5745-5755, 2007.
23. H. Tagawa and M. Seto, A microRNA cluster as a target of genomic amplification in malignant lymphoma, Leukemia, 19, 2013-2016, 2005.
24. L. He, J. M. Thomson, M. T. Hemann, E. Hernando-Mongre, D. Mu, S. Goodson, S. Powers, C. Cordon-Cardo, S. W. Lowe, G.J. Hannon and S.M. Hammond, A microRNA as a potential human oncogene, Nature, 435, 828-833, 2005.

25. C. Xiao, L. Srinivasan, D. P. Calado, H. C. Patterson, B. Zhang, J. Wang, J.M. Henderson, J. L. Kutok and K. Rajewsky, Lymphoproliferative disease and autoimmunity in mice with increased miR-17-92 expression in lymphocytes, Nature Immunol., 9, 405-414, 2008.
26. A. Ventura, A.G. Young, M. W. Winslow, L. Lintault, A. Messner, S.J. Erkeland, J. Newman, R.T. Bronson, D. Crowley, J.R. Stone, R. Jaenisch, P.A. Sharp and T. Jacks, Targeted deletion reveals essential and overlapping functions of the miR-17 through 92 family of miRNA clusters, Cell, 132, 875-886, 2008.
27. P. van Vlierberghe, I. Homminga, L. Zuurbier, J. Gladdines-Buijs, E.R. van Wering, M. Horstmann, HH.B. Beverloo, R. Pieters and J.P.P. Meijerink, Cooperative genetic defects in TLX3 rearranged pediatric T-ALL, Leukemia, 22, 762-770, 2008.
28. E. Ruiz-Ballesteros, M. Mollejo, M. Mateo, P. Algara, P. Martinez and M.A. Piris, MicroRNA losses in the frequently deleted region of 7q in SMZL, Leukemia, 21, 2547-2549, 2007.
29. S. Varambally, Q. Cao, R-S. Mani, S. Shankar, X. Wang, B. Ateeq, B. Laxman, X. Cao, et al., Genomic loss of microRNA-101 leads to overexpresssion of histone methyltransferase EZH2 in cancer, Science, 322, 1695-1699, 2008.
30. H. Yamada, K. Yanagisawa, S. Tokumaru, A. Taguchi, Y. Nimura, H. Osada, M. Naagino and T. Takahashi, Detailed characterization of a homozygously deleted region corresponding to a candidate tumor suppressor locus at 21q11-21 in human lung cancer, Genes, Chromosomes Cancer, 47, 810-818, 2008.
31. K. Nagayama, T. Kohno, M. Sato, Y. Arai, J.D. Minna and J. Yakota, Homozygous deletion scanning of the lung cancer genome at a 100-kb resolution, Genes, Chromosomes Cancer, 46, 1000-1010, 2007.
32. S. Pfister, M. Remke, M. Castoldi, A. H. C. Bai, M.U. Muckenthaler, A. Kulozik, A. von Deimling, A. Pscherer, P. Lichter, A. Korshunov, Novel genomic amplification targeting thee microRNA cluster at 19q13.42 in a pediatric embryonal tumor with abundant neuropil and true rosettes, Acta Neuropathol Dec 5, 2008 [Epub ahead of print]
33. I.V. Makunin, M. Pheasant, C. Simons and J.S. Mattick, Orthologous microRNA genes are located in cancer-associated genomic regions in human and mouse, PLoS One, 11, e1133, 2007.
34. C. Slape, H. Hartung, Y-W. Lin, J. Bies, L. Wolff and P.D. Aplan, Retroviral insertional mutagenesis identifies genes that collaborate with NUP98-HOXD13 during leukemic transformation, Cancer Res., 67, 5148-5155, 2007.
35. C. L. Wang, B. B. Wang, G. Bartha, L. Li, N. Channa, M. Klinger, N. Killeen and M. Wabl, Activation of an oncogenic microRNA cistron by provirus integration, Proc. Natl. Acad. Sci. USA, 103, 18680-18684, 2006.
36. A.M. Lum, B. B. Wang, L. Li, N. Channa, G. Bartha and M. Wabl, Retroviral activation of the mir-106a microRNA cistron in T lymphoma, Retrovirology, 4, 5 (2007).
37. A. Donsante, D.G. Miller, Y. Li, C. Vogler, E.M. Brunt, D.W. Russell and M.S. Sands, AAV vector integration sites in mouse hepatocellular carcinoma, Science, 317, 477 (2007).
38. T. Suzuki, K. Minehata, K. Akagi, N.A. Jenkins and N.G. Copeland, Tumor suppressor gene identification using retroviral insertional mutagenesis in Blm-deficient mice, EMBO J., 25, 3422-3431, 2006.
39. V. Theodorou, M.A. Kimm, M. Boer, L. Wessels, W. Theelen, J. Jonkers and J. Hilkens, MMTV insertional mutagenesis identifies genes, gene families and pathways involved in mammary cancer, Nat. Genet. 39, 759-769, 2007.

40. M. Sauvageau, M. Miller, S. Lemieux, J. Lessard, J. Hebert and G. Sauvageau, Quantitative expression profiling guided by common retroviral insertion sites reveals novel and cell type specific cancer genes in leukemia, Blood, 111,790, 2008.
41. F.K. Johansson, J. Brodd, C. Ekløf, M. Ferletta, G. Hesselager, C.F. Tiger, L. Uhrbom and B. Westermark, Identification of candidate cancer-causing genes in mouse brain tumors by retroviral tagging, Proc. Natl Acad Sci. USA, 101,11334-11337, 2004.
42. M. Yanagida, M. Osato, N. Yamashita, H. Liqun, B. Jacob, F. Wu, X. Cao, T. Nakamura, T. Yokomizo, S. Takahashi, M. Yamamoto, K. Shigesada and Y. Ito, Increased dosage of Runx1/AML1 acts as a positive modulator of myeloid leukemogenesis in BXH2 mice, Oncogene, 24, 4477-4485, 2005.
43. T. Suzuki, H. Shen, K. Akagi, H.C. Morse, J.D. Malley, D.Q. Naiman, N. A. Jenkins and N.G. Copeland, New genes involved in cancer identified by retroviral tagging, Nat. Genet., 32, 166-174, 2002.
44. H. Mikkers, J. Allen, P. Knipscheer, L. Romeijn, A. Hart, E. Vink and A. Berns, High-throughput retroviral tagging to identify components of specific signaling pathways in cancer, Nat. Genet., 32, 153-159, 2002.
45. A.H. Lund, G. Turner, A. Trubetskov, E. Verhoeven, E. Wientjens, D. Hulsman, R. Russell, R.A. DePinho, J. Lenz and M. van Lohuizen, Genome-wide retroviral insertional tagging of genes involved in cancer in Cdkn2a-deficient mice, Nat. Genet., 32,160-165, 2002.
46. J. Bijl, M. Sauvageau, A. Thompson and G. Sauvageau, High incidence of proviral integrations in the Hoxa locus in a new model of E2a-PBX1-induced B-cell leukemia, Genes Dev., 19,224-233, 2005.
47. N. Yamashita, M. Osato, L. Huang, M. Yanagida, S.C. Kogan, M. Iwasaki, T. Nakamura, K. Shigesada, N. Asou and Y. Ito, Haploinsufficiency of Runx1/AML1 promotes myeloid features and leukaemogenesis in BXH2 mice, Br. J. Haematol., 131, 495-507, 2005.
48. H.C. Hwang, C.P. Martins, Y. Bronkhorst, E. Randel, A. Berns, M. Fero and B.E. Clurman, Identification of oncogenes collaborating with p27Kip1 loss by insertional mutagenesis and high-throughput insertion site analysis, Proc Natl Acad Sci. USA 99,11293-11298, 2002.
49. M. Graham, J.M. Adams and S. Cory, Murine T lymphomas with retroviral inserts in the chromosomal 15 locus for plasmacytoma variant translocations, Nature, 314, 740-743, 1985.
50. J-C. Cui, Y-J. Li, A. Sarkar, J. Brown, Y-H. Tan, M. Premyslova, C. Michaud, N. Iscove, G-J. Wang and Y. Ben-David, Retroviral insertional activation of the Fli-3 locus in erythroleukemias encoding a cluster of microRNAs that convert Epo-induced differentiation to proliferation, Blood, 110,2631-2640, 2007.
51. H. Schwindt, T. Akasaka, R. Zühlke-Jenisch, V. Hans, C. Schaller, W. Klapper, M.J. Dyer, R. Siebert and M. Deckert, Chromosomal translocations fusing the BCL6 gene to different partner loci are recurrent in primary central nervous system lymphoma and may be associated with aberrant somatic hypermutation or defective class switch recombination, J. Neuropathol. Exp. Neurol., 65,776-782, 2006.
52. C. Mayr, M.T. Hemann and D.P. Bartel, Disrupting the pairing between let-7 and Hmga2 enhances oncogenic transformation, Science, 315, 15761579, 2007.
53. W.Y. Chen, X. Zeng, M.G. Carter, C. N. Morrell, R-W. C. Yen, M. Esteller, D.N. Watkins, J.G. Herman, J.L. Mankowski and S.B. Baylin, Heterozygous disruption of Hic1 predisposes mice to a gender-dependent spectrum of malignant tumors, Nat. Genet., 33,197-202, 2003.

CHAPTER 8

INTERCONNECTION OF MICRORNA AND TRANSCRIPTION FACTOR-MEDIATED REGULATORY NETWORKS

Yiming Zhou

*Myeloma Institute for Research and Therapy,
University of Arkansas for Medical Sciences,
4301 West Markham Street
Little Rock, AR 72205*

MicroRNA, a class of newly discovered non-coding small RNAs with mature sequences containing only ~22 nucleotides, adds an additional layer of complexity to the cellular regulatory network. Like other players in this network, microRNAs perform their regulatory functions by closely interacting with transcription factors and other regulatory proteins. The expression of miRNAs is controlled by transcription factors. In turn, microRNAs repress the translation of thousands of genes including transcription factors. microRNAs significantly affect many types of cellular processes. Furthermore, in most cases, microRNAs and transcription factors work synergistically to regulate their common targets. These interactions form a microRNA-mediated regulatory network. Understanding the interactions is critical to fully understand the functions of microRNAs in context of cellular regulatory networks. With the advances in our understanding of how microRNAs are regulated and how miRNAs regulate their targets, researchers are getting opportunities to decipher the complexity of interactions between miRNAs and transcription factors. Here we review the state of the art in research of the interconnections of miRNA- and transcription factor-mediated networks.

1. Introduction

MicroRNAs (miRNAs) belong to a class of non-coding small RNAs with mature sequences containing ~22 nucleotides. The first miRNA lin-4 was found in *C. elegans*. lin-4 controls the timing of *C. elegans* larval development.[1-3] Lee *et al.*

and Wightman et al.[2,3] first recognized that miRNAs regulate protein expression at a post-transcriptional stage. So far, no miRNA is found in single-cell organisms, except a green alga Chlamydomonas reinhardtii.[4,5] Since then the properties of this novel and vital class of regulators are being extensively studied. Although some miRNAs may be transcribed by RNA polymerase III (Pol III),[6] it is believed that most miRNAs are transcribed by RNA polymerase II (Pol II).[7] In mammalian cells, the primary transcripts (pri-miRNA) are cleaved in cell nucleus by the Drosha into 60~70nt precursors with stem-loop structure (pre-miRNA).[8] The pre-miRNAs are then exported to cytoplasm by Exportin-5 and its cofactor Ran-GTP.[9-11] Finally, Dicer crops the exported miRNAs in the cytoplasm into ~22nt mature miRNAs.[12,13] Mature miRNAs are eventually transferred to Argonaute proteins and serve as guides in mRNA silencing.[14] Expression studies have shown that many miRNAs have tissue-specific or developmental-stage-specific expression patterns.[7,15]

So far most studies showed that in vertebrates the main role of miRNAs is to repress the gene expression by inhibiting protein translation initiation with or without directly degrading mRNAs. But a few recent studies also reported that miRNAs may enhance mRNA translation within specific context.[16-18] Since more than one-third of human genes appear to have been under selective pressure to maintain their pairing to miRNA seeds during evolution,[19] it not surprising that emerging evidences from *in vivo* and *in vitro* experiments are showing that miRNAs regulate a broad diversity of cellular processes, including differentiation, development, aging, apoptosis, metabolism and oncogenesis.[7,20-24]

Cellular regulatory network is a network representing how the genes in a cell are being regulated and orchestrated to respond to outside stimuli or execute intracellular programs. The network is formed by genes (nodes) and the regulatory relationship between genes (connections). Previously, transcription factors (TF) were the predominant regulators in the network. And the interactions within the transcription factor networks have been studied for decades. These studies were boosted recently by availability of high-throughput experimental methods like ChIP-chip, which combines chromatin immunoprecipitation ("ChIP") techniques with the DNA microarray technology ("chip"). The new players, miRNAs, confer the current known regulatory network of gene expression with an additional layer, indicating that the network is even more complex than expected before. Here we will discuss how the miRNA can be integrated into the currently known transcription regulation network to improve our understanding of the regulation of mammalian cells.

2. Targets of miRNAs

Mature miRNAs is first linked to RNA-induced silencing complex (RISC) and then bound to target mRNAs to block the translation of the mRNAs or degrade them by yet unclear mechanisms.

Comprehensive understanding of the underlying mechanisms of miRNA-mediated repression is lacking. Efficient high-throughput experimental methods for miRNA target identification are still underdeveloped. Therefore, genome-wide identification of miRNA targets is currently based on computational predictive models.

In plants, a miRNA target typically contains a sequence that is complementary to the sequence of the whole miRNA.[7] The complete match of ~22 nucleotides is long enough to guarantee that there is only one target in whole genome for most miRNAs. Thus the predictions of miRNA targets are straightforward in plant. Interestingly, most targets of miRNAs in plants are TFs,[7,25] which give hints for functions of miRNAs.

In vertebrates, the situation is more complex since a perfect complementarity is not necessary for a miRNA to recognize its targets. So it is almost impossible to identify targets of miRNAs in vertebrate genomes by simple sequence comparison methods. It was first learnt from as few as 3 miRNAs and their experimentally validated targets that a seed region at the 5' of miRNAs should approximately match with the 3' untranslated regions (UTR) of the putative target genes, while 3' of miRNAs is less important.[3,26-28] A special mismatch G:U wobble, which is thermodynamically favorable and common in RNA secondary structure, is usually considered as detrimental in seed region while maybe acceptable in non-seed region.[29] So far, perfect Watson-Crick complementarity to bases 2–8 of mature miRNA sequences (numbered from the 5') is the most useful and comprehensively applied feature for targets identification in miRNAs. However, this single feature is not always sufficient for correctly predicting the targets of miRNAs.[30]

Since repression function of miRNAs is essentially executed by basepairing between sequences of miRNAs and targets, it is supposed that a low folding free energy ΔG of RNA–RNA duplexes is energetically favorable. By applying this additional rule, the information about 3' non-seed region of miRNA sequences is used to contribute to specificity of miRNA target selection.

Another important information for identifying targets of miRNAs is the conservation of miRNA binding sites at 3' UTRs across species. The hypothesis is that a functional binding site should be kept as same as possible during the evolution to avoid loss of function. Several lines of evidence support this hypothesis. First, the 5' ends of related miRNAs tend to be better conserved than the 3' ends.[31] This finding also supported the hypothesis that 5' region of miRNAs is most critical for mRNA recognition. Second, signal : noise ratio for prediction of targets of miRNAs increases with considering conservation of binding sites across two or more species. Interestingly, without using any miRNA information and only focusing on the conserved sequences in 3' UTR of human, mouse, rat and dog genomes, Xie *et al.* created a systematic catalogue of common regulatory motifs in 3' UTRs, in which nearly one-half are associated with miRNAs.[32] This highlighted the importance of conservation information for identifying the targets of miRNAs. Notably, while conservation is useful to reduce the signal : noise ratio of prediction, it also greatly reduced the number of predicted targets of miRNAs (by ~5-fold[28]) and therefore increased false negative results.

In many cases, a single binding site at 3' UTR is not sufficient to induce the repression. There are two conserved binding sites of let-7 at 3' UTR of lin-41 in *C. elegans*. Vella *et al.* found that the two binding sites are necessary and sufficient for proper lin-41 down-regulation.[33] Even in the case that one binding site is sufficient to repression translation, additional binding sites at 3'UTR of target genes can enhance the repression effect.[29] Of note, the spacer sequence between two binding sites is also important for the down-regulation activity.[29,33] The observation that target sites for distinct miRNAs are often simultaneously conserved, further suggested combinatorial regulation by multiple miRNAs.[34]

The importance of spacer sequences may reflect the necessity of target-site accessibility, as determined by base-pairing interactions within the mRNA, in miRNA target recognition.[35-37] Vella *et al.*[38] and Grimson *et al.*[39] also recognized that context of miRNA binding site can be critical for function. Considering that sequences of mRNAs can have second-structure, in which some sites could be closed to even matched miRNAs, a model suggested a two-step hybridization reaction during reorganization of a miRNA and its target sequence: nucleation at an accessible target site followed by hybrid elongation to disrupt local target secondary structure and form the complete miRNA-target duplex.[36] The site accessibility could be as important as sequence match in the seed for determining the efficacy of miRNA-mediated translational repression.[35]

Table 1 listed some of the most commonly used algorithms to predict the targets of vertebrate miRNAs, including PicTar,[40] TargetScan,[19,27] DIANA-microT,[41] RNAhybrid,[42] miRBase targets,[43] mirWIP,[30] PITA.[35] These algorithms employ different combination of features to determine whether a sequence is a target of miRNAs. Although most of algorithms used simple rule-based procedure to train model and predict the targets of miRNAs, a variety of machine learning algorithm had also been applied in miRNA targets predictions, such as SVM[44] and random walk.[45] It is not clear whether these sophisticated methods perform much better than others.

Table 1. Several algorithms for prediction of targets of miRNAs and the features used.

Algorithms	seed region	ΔG	site accessibility	conservation	position in 3'UTR	G:U penalty
TargetScan	1	1	1	1	1	1
mirWIP	1	1	1	1	0	1
PITA	1	1	1	1	0	1
DIANA-microT	1	1	0	1	0	1
RNAhybrid	1	1	1	1	0	0
PicTar	1	1	0	1	0	1
miRBase targets	1	1	0	1	0	1

Despite the lack of sufficient numbers of experimentally verified targets for accurate assessment of the overall sensitivity and specificity of the predictions obtained by these algorithms, some reports indicated that a large class of miRNA targets can be confidently predicted.[46,47] However, computational approaches for identification of miRNA targets still have significant limitations. Statistical analysis indicates that even the best predicted target sites are at the border of significance; thus, target predictions should be considered as tentative until experimentally validated.[28] Since performance of any single algorithm is not completely satisfying, it is intuitive to utilize the intersection or union of all predicted targets from a variety of distinct algorithms. But the issue with this approach is that union provides too much targets in which substantial amount is false positive, while intersection provides too few targets in which substantial amount of true targets is missed (false negative). Sethupathy et al. provided an approach to how to combine the different algorithm to achieve higher quality of predicted targets of miRNAs.[47]

The difficulty of the prediction may come from the fact that RISC complex utilizes a variety of information to find the paired sequences of miRNAs, like demonstrated by Grimson et al.[39] In this study, authors listed many features that could affect the miRNA-mediated repression of protein expression, including base pairing beyond seed region, local AU composition within 3'UTRs or flanking the binding sites, relative position of binding sites within 3'UTR of target genes. The difficulty may also come from the repression effect of miRNAs, which is not binary. The repression effect might be continuous and the cooperation of miRNA binding sites might greatly affect the effect of miRNA-mediated repression. So it may not be sufficient to simply determine whether a gene is the target of a miRNA or not.

Fortunately, platforms for determining the targets of miRNAs at a large scale are now emerging. A few groups had reported the systems utilizing the antibody of components of RISC, a complex that miRNAs bound to during interaction with mRNAs, to pull out the miRNAs and their bound targets in vitro and in vivo.[48-50] The pull-out target sequences can be then identified through microarray hybridization. The positive correlation between the ectopically overexpressed miRNA and RISC-bound mRNAs provided a useful way to determine targets for specific miRNAs.[48] Baek et al.[51] and Selbach et al.[52] recently developed mass spectrometry-based methods that determine the genuine targets of miRNAs, by checking the alteration of protein expression level on a whole proteome scale.

The paired gene expression profiles and miRNA expression profiles were used to determine hundreds of targets of up-regulated or down-regulated miRNAs by simply checking the correlation coefficients of gene expression level and miRNA expression level. This type of approach can not identify the relationship between miRNAs and their targets that were only blocked for translation without degradation of mRNAs.[53]

In the future, both computational approaches and experimental approaches will be required for identifying the targets of miRNAs. Although experimental approaches generally produce more reliable targets, its application was limited by the cost and the dynamic properties of cells, namely, at any specific spatio-temporal context, experimental approaches can only determine a part of miRNA: target relationships. With computational approach it is possible to predict the targets regardless of context. With emerging of the more experimentally validated targets, the principles of miRNA-target binding will be better understood and will promote better prediction algorithms.

3. Regulators of miRNAs

As most of miRNAs are transcribed by RNA polymerase II (Pol II),[7] the major regulators of expression of miRNAs are TFs. Clarifying the regulators of miRNAs is critical for fully understanding the miRNA-mediated regulatory networks. The high-throughput ways to determine the TFs of given genes is TF binding matrix (motif)-based methods. Transcription-factor binding sites (TFBSs) are usually short (around 5-15 base-pairs) and frequently degenerate sequences. The pattern of these degenerate sequence is presented by a matrix (motif), giving the probability distribution of A, C, T and G at each site. The motifs of TFs were summarized from known TFBSs, which were determined by a variety of experimental approaches. Transfac[54] and Jaspar[55] are two major collections of currently annotated TF binding motifs. Predicted TF-target relationships can be determined by scanning promoter regions of given genes with motifs. ChIP-based methods can experimentally determine the binding sites for a given TF on a large scale. But high-throughput experimental methods to determine all possible TFs for given genes on a large scale remain in demand.

One of most challenging issues for identifying transcription regulators of miRNAs is that the promoter regions of miRNAs are largely unknown since only a minority of the primary transcripts have been identified and annotated. Prior attempts to identify transcription regulators of miRNAs have searched for TF binding sites only close to the annotated mature miRNA sequences.[56-58] Most recently, a few studies tried to define the promoter regions of miRNAs by taking advantages of chromatin signatures at the transcriptional start sites of most genes in the genome.[59,60] The studies shows that more than half of all investigated miRNAs have transcription start sites (TSS) that are more than 5,000 basepairs away from mature miRNAs. More accurate determination of the promoter regions of miRNAs will greatly improve the quality of predicted transcription regulators of miRNAs.

More detailed discussion about identifying TFBS is beyond the scope of this review and has been comprehensively reviewed elsewhere.[61,62]

4. Inter- and Intra-Combinatorial Regulations by Transcription Factors and miRNAs

The sole function of miRNAs is regulation of translation from mRNA to protein. As newly discovered players in gene expression regulatory network, miRNAs add an additional level to already complex gene expression regulatory network. Like other players in this network, miRNAs perform their regulatory functions by closely interacting with others. These interactions form a miRNA-mediated network. Understanding the network is critical to fully understand the functions of miRNAs and cellular regulatory networks.

The theoretical studies of the networks are carried out at two levels. The fundamental one is to discover the structure of the network. Based on the structure, the functions and properties of the network can be then deciphered. The structure of the network is built by the nodes and connections between nodes. The nodes could represent miRNAs, TFs, target genes of miRNAs and TFs and regulators of miRNAs. The connections represent the regulatory relationships between the nodes. We have discussed approaches used to discover the regulatory relationships between miRNAs and their targets or regulators above.

The human genome has ~25,000 genes, ~2000 TFs[63] and more than 700 miRNAs (http://microrna.sanger.ac.uk). The network built on so many nodes is extremely complex. In spite of the recent advances of computational biology and large-scale experimental techniques, studying the genome-wide network as a whole remains unfeasible. Fortunately, researchers have found universal features of structure of such complex networks, such as scale free/hub topology[64] and modularity/hierarchical topology.[65,66] In a scale free topology, the number of connections of a given node follows a power law, namely a minority of nodes in the network has a majority of connections while a majority of nodes has a minority of connections (Fig. 1). The minority of genes with lot of connections are called "hubs". A module is a subset of the network, in which a clique of nodes sharing a common function is highly mutually inter-connected, while there are much fewer connections between them and outsider nodes. The modules are then connected into bigger networks with a hierarchical structure. With the modular/hierarchical structure, cells have acquired the means to control the input of a system in a specific module, without spreading it everywhere.

Another significant finding is that a big network can be broken down into small sub-networks, such as two-component, three-component sub-networks. These sub-networks can be seen as building blocks of the whole network and their functions and properties can be integrated to infer the functions and properties of larger networks formed by them. Furthermore, by simply counting the occurrence of the sub-network, researchers found that some types of topologies of sub-networks appeared at much higher frequency than others. Those frequently occurred sub-networks are called "motifs".[67] One of the advantages of using the concept of motif is the properties of motifs are not dependant on specific types of molecules but are defined by the topology of connections/interactions. So the comprehensive study of limited motifs helps to understand properties of larger networks configuration.

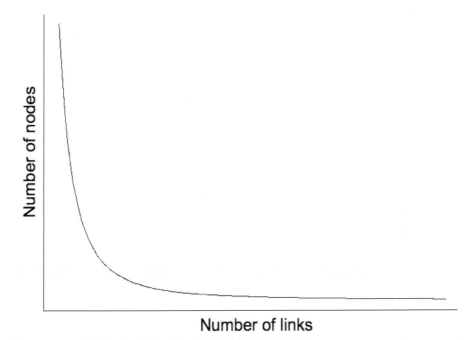

Figure 1. In a scale free topology, the number of connections of a given node follows a power law, namely, a minority of nodes in the network get a majority of connections while a majority of nodes a minority of connections.

4.1. *Intra-combinatorial regulations by miRNAs*

Since there is no existing evidence that a miRNA can directly regulate another miRNA, a pure miRNA regulatory network is quite simple, in which two or more miRNAs co-regulate a target (Fig. 2). There are multiple lines of evidence showing that the cooperation between miRNAs is universal and critical for their functions. By investigating predicted miRNA targets, several groups found that multiple miRNAs prefer to bind to a single target at same time.[29,40,57,68,69] Moreover, target sites for distinct miRNAs are often simultaneously conserved, suggesting combinatorial regulation by multiple miRNAs.[34] Recent large-scale knockout miRNA study, which showed most *C. elegans* miRNAs are individually not essential for development or viability,[70] further highlighted the importance of cooperation of miRNA binding sites.

For the structure of the pure miRNA regulatory network, the most important feature is the type of cooperation between miRNAs. There are two type of cooperation, AND-gate or OR-gate (Fig. 2). AND-gate means that the repression of target only occurs when multiple miRNAs bind to the target at the same time.

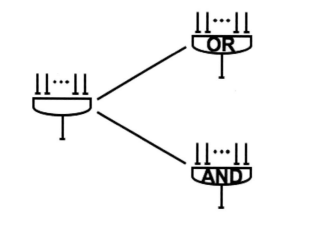

Figure 2. Two or more miRNAs co-regulate a target by an OR-gate or an AND-gate.

OR-gate means repression can occur when one or more miRNAs bind to the target. For the OR-gate, we had to further determine if the cooperation is an additive model or a multiplicative model. In an additive model, the effect of all miRNA binding sites for a single target is equal to the sum of effects of each binding site. In a multiplicative model, the effect of all miRNA binding sites for a single target is greater than the sum of effects of each binding site. In many cases, the effect of different binding sites of a same miRNA is equal to or less than sum of the effect of different binding sites of a same miRNA.[29,51,71] This indicates that the cooperation of these miRNA binding sites is more likely to be additive OR-gate. However, when the two sites were close together, the repression tended to be greater than that expected from the independent contributions of two single sites, a multiplicative OR-gate.[39] This observation suggested that the type of cooperation of miRNA binding site might be determined by the distances between two sites.

Shalgi et al. revealed several levels of hierarchy in an intra-miRNA network, whereby a few miRNAs interact with many other weakly connected miRNA partners.[57] Examination of the degree distribution of each miRNA further revealed that the network is scale-free.[57]

The cooperation between miRNAs can have a special effect in miRNAs' functions. A miRNA can bind to hundreds of distinct mRNA targets. The synergy of different miRNAs can provide a higher specificity. The specificity of miRNAs might be achieved by the cooperation. The targets with only one miRNA binding site can be weakly repressed. Only the targets with multiple multiplicative-type binding sites can be significantly repressed. In addition, by switching its partner miRNAs, a miRNA could switch their specificity temporally and spatially.

4.2. Inter-interactions between miRNAs and TFs

Studies had demonstrated that the TFs and miRNAs extensively interact with each other.[56,57,69] Not like miRNAs in plants, which predominantly target TFs[7,25] miRNAs in mammals can target thousands of targets involving a whole spectrum of genes. But the predicted regulatory targets of mammalian miRNAs are also enriched for genes involved in transcriptional regulation.[27,56] Of note, although miRNAs and TFs are extensively interacting, overall the combinatorial regulation by TF-miRNA pairs is less abundant than combinatorial regulation by TF pairs and miRNA pairs, perhaps due to a higher probability of evolutionary duplication events for shorter DNA sequences.[69]

Gene enrichment analysis showed that functionally related genes are reciprocally regulated by strongly interacting pairs of TF-miRNA.[69] Interestingly, genes associated with cell adhesion-related GO terms including homophilic cell adhesion, cell-cell adhesion, and the less specific cell adhesion category, were most likely to be co-regulated by TF-miRNA pairs.[69] This observation is not only consistent with the current presumption that miRNAs are only found in multi-cellular species, but also provides an important clue for their cellular roles and origin. Moreover, since the reduction in cell adhesion correlates with tumor invasion, it implicates involvement of these miRNAs with cancer metastasis. We found that 8 out of the 40 cell adhesion-enriched miRNAs are present in a dataset of 21 miRNAs that are up or down regulated in a variety of cancers.[69]

4.2.1. Structure of miRNA-TF co-regulatory network

A number of recent studies has revealed that "target hub" genes bearing massive TF-binding sites were potentially subject to massive regulation by dozens of miRNAs.[57,58] The findings are in line with a hypothesis that nodes with more connections will more likely get new connections during the evolution. The top genes with large number of both TFBS and miRNA binding sites were enriched in the functions associated with development, which indicated that this kind of fine-tuning is important for precisely controlling the development and differentiation of cells. Many of these "target hub" genes are transcription regulators.[57] This suggested that it is an important pathway for miRNAs to indirectly regulate genes through repressing TFs.

MiRNAs could be "target hub" genes too. A study found a class of miRNAs that is regulated by a large number of TFs, whereas the others are regulated by only a few TFs.[56] Comparing miRNA expression pattern in embryonic developmental stages and adult tissues or cancer samples reveals that the first class of miRNAs are higher expressed in embryonic developmental stages, while the second class more highly expressed in adult tissues or cancer samples.[56]

In addition, "regulator hub" genes, which regulate massive number of targets, are more likely to have interactions with miRNAs. Cui *et al.* revealed that in a signaling pathway, when an adaptor has potential to recruit more downstream components, these components are more frequently targeted by miRNAs.[72] It was also found that miRNAs prefer to co-regulate the targets with master TFs that might regulate thousands of genes.[69] "Regulator hub" genes are Achilles' heel of the complex network, since perturbations on these genes might disrupt functions of too many target genes. As miRNAs had been proposed to buffer stochastic perturbations,[73] the preference of miRNAs to "regulator hub" could confer robustness to the regulatory network.

4.2.2. *Motifs*

Motifs are a class of recurrently appeared sub-structure in a network, which might be building blocks of networks. Different motifs have a different function and response to different stimulation or the execution of the inner cell program.

The simplest cooperation between a miRNA and a TF is to co-regulate a target without interaction between them. This motif is the second over-represented motif in 46 possible configurations with three components.[56] By this structure, miRNAs can work as a buffer of TF's activity[73] or add specificity to a widely expressed TF.[74]

The most recurrently appeared motif in a variety of networks is a feed forward loop (FFL)[67,75,76] A diversity of feed-forward loops is also over-represented in miRNA-mediated regulatory networks.[56,57] A type of FFL, in which two TFs regulate each other and one miRNA regulates both of the factors (Fig. 3A), is a third over-represented motif.[56] Mathematical modeling shows that this motif may function as resisting environmental perturbation, providing one mechanism contributed by miRNAs to explain the robustness of developmental programs that.[56] Another type of FFL, in which a TF activates its targets and a miRNA represses both TFs and targets of the TF (Fig. 3B), was over-represented

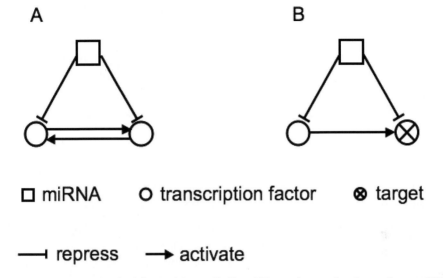

Figure 3. Two types of a feed-forward loop. A) Two TFs regulate each other and one miRNA regulates both of the factors. B) A TF activates its targets and a miRNA represses both TFs and targets of the TF.

when the miRNA and TF shared significant common targets.[69] Generally, miRNAs could repress hundreds of mRNAs, which in turn could be wastefully produced if the relevant TFs do not change their level of activity. This type of FFL can avoid this inefficiency and accelerate mRNA degradation by simultaneously suppress TF's target and TF itself.

As additional regulators, one interesting question is what type of motif in intra-TF network is more likely targeted by miRNAs. Cui et al. found miRNAs predominantly target positive regulatory motifs, highly connected scaffolds and most downstream network components such as signaling TFs, but less frequently target negative regulatory motifs, common components of basic cellular machines and most upstream network components such as ligands.[72]

Although they were not identified as top recurrent motifs, some of the motifs had been reported to play critical and interesting roles in regulation of differentiation and other conditions. A double-negative feedback loop involving miR-273 provide a bistable switch that decide a neuronal cell asymmetric fate, the stable ASEL or ASER terminal end state.[77] Using a computational approach, Wu and Xie identified a double-negative feedback loop between REST and the miRNAs (miR-124a, miR-9 and miR-132) in stabilizing and maintaining neuronal gene expression.[78]

4.3. *miRNA and protein-protein interaction network*

Almost all cellular processes are performed by proteins. Interactions between proteins provide a mechanistic basis for regulation of most of biological processes in an organism. Since the functional state of a protein-protein interaction network depends on gene expression, an interesting question is what relationships exist between protein interaction network and miRNA-mediated regulation.

Liang and Li showed that protein connectivity in the human protein-protein interaction network is positively correlated with the number of miRNA target-site types in the 3' untranslated regions of the gene encoding the protein, and that interacting proteins tend to share more miRNA target-site types than random pairs.[79] Moreover, their results demonstrate that miRNA-targeting propensity for genes in different biological processes can be largely explained by their protein connectivity. Finally, they showed that for hub proteins, miRNA regulation complexity is negatively correlated with clustering coefficient, suggesting that miRNA regulation is more important to inter-modular hubs than to intramodular hubs. Clustering coefficient indicates the extent to which the neighbors of a given node in a network are inter-connected.[80]

Hsu *et al.* performed topological analysis to elucidate the global correlation between miRNA regulation and protein-protein interaction network in human. The analysis showed that target genes of individual miRNA tend to be hubs and bottlenecks in the network. While proteins directly regulated by miRNA might not form a network module themselves, the miRNA-target genes and their interacting neighbors jointly showed significantly higher modularity.[81]

Focusing on a TP53-associated network with higher data quality, 114 of these 141 p53-interactants were predicted to be regulated by p53-miRNAs and were shown to be statistically significant ($p = 1.41e-21$).[82]

Overall, these studies revealed the importance of the miRNA-mediated regulation of the protein interaction network.

5. Conclusions and the Future

With currently available data, we have achieved significant advances in understanding miRNA-mediated regulatory network, for instance module/hub structure, FFL as most recurrently appeared motif and regulation of the protein-protein interaction network. Since the findings were independently revealed by different groups using different data sets, they are quite robust, in spite of the great unavoidable noise in the data currently available. Nevertheless, it is imaginable that much more significant advances will be achieved with availability of data with higher quality and more comprehensive analysis.

First, the newly developed high-throughput techniques for experimentally identifying the targets of miRNA are revealing a huge amount of reliable miRNA-target relationships, which will in turn help to develop novel algorithms for better target prediction of miRNAs. The high-throughput techniques for determining the promoter regions of miRNAs will soon be used to improve the quality of the prediction of regulators of miRNAs. Thus these novel techniques are conferring us better understanding for the structure of miRNA-mediated regulatory network.

Second, with more and more miRNA profiles available in different tissues, different developmental stages and different diseases, the patterns of miRNA expression will be discovered. Furthermore, coupling these miRNA expression profiles with mRNA expression profiles, we will be able to decipher the dynamics of the miRNA-mediated regulatory network in different context.

Finally, the functional relevance of the networks in vitro or in vivo remains to be explored. And the applications of principles of miRNA-mediated regulatory network in translational biomedicine are highly in demand.

References

1. Ambros V, Horvitz HR: Heterochronic mutants of the nematode Caenorhabditis elegans. Science 1984, 226(4673):409-416.
2. Lee RC, Feinbaum RL, Ambros V: The C. elegans heterochronic gene lin-4 encodes small RNAs with antisense complementarity to lin-14. Cell 1993, 75(5):843-854.
3. Wightman B, Ha I, Ruvkun G: Posttranscriptional regulation of the heterochronic gene lin-14 by lin-4 mediates temporal pattern formation in C. elegans. Cell 1993, 75(5):855-862.
4. Molnar A, Schwach F, Studholme DJ, Thuenemann EC, Baulcombe DC: miRNAs control gene expression in the single-cell alga Chlamydomonas reinhardtii. Nature 2007, 447(7148):1126-1129.
5. Zhao T, Li G, Mi S, Li S, Hannon GJ, Wang XJ, Qi Y: A complex system of small RNAs in the unicellular green alga Chlamydomonas reinhardtii. Genes & development 2007, 21(10):1190-1203.
6. Borchert GM, Lanier W, Davidson BL: RNA polymerase III transcribes human microRNAs. Nature structural & molecular biology 2006, 13(12):1097-1101.
7. Bartel DP: MicroRNAs: genomics, biogenesis, mechanism, and function. Cell 2004, 116(2):281-297.
8. Lee Y, Ahn C, Han J, Choi H, Kim J, Yim J, Lee J, Provost P, Radmark O, Kim S et al: The nuclear RNase III Drosha initiates microRNA processing. Nature 2003, 425(6956):415-419.
9. Bohnsack MT, Czaplinski K, Gorlich D: Exportin 5 is a RanGTP-dependent dsRNA-binding protein that mediates nuclear export of pre-miRNAs. RNA 2004, 10(2):185-191.
10. Kim VN: MicroRNA precursors in motion: exportin-5 mediates their nuclear export. Trends in cell biology 2004, 14(4):156-159.

11. Yi R, Qin Y, Macara IG, Cullen BR: Exportin-5 mediates the nuclear export of pre-microRNAs and short hairpin RNAs. Genes & development 2003, 17(24):3011-3016.
12. Hutvagner G, McLachlan J, Pasquinelli AE, Balint E, Tuschl T, Zamore PD: A cellular function for the RNA-interference enzyme Dicer in the maturation of the let-7 small temporal RNA. Science 2001, 293(5531):834-838.
13. Knight SW, Bass BL: A role for the RNase III enzyme DCR-1 in RNA interference and germ line development in Caenorhabditis elegans. Science 2001, 293(5538):2269-2271.
14. Hutvagner G, Zamore PD: A microRNA in a multiple-turnover RNAi enzyme complex. Science 2002, 297(5589):2056-2060.
15. Zeng Y: Principles of micro-RNA production and maturation. Oncogene 2006, 25(46):6156-6162.
16. Henke JI, Goergen D, Zheng J, Song Y, Schuttler CG, Fehr C, Junemann C, Niepmann M: microRNA-122 stimulates translation of hepatitis C virus RNA. The EMBO journal 2008, 27(24):3300-3310.
17. Orom UA, Nielsen FC, Lund AH: MicroRNA-10a binds the 5'UTR of ribosomal protein mRNAs and enhances their translation. Molecular cell 2008, 30(4):460-471.
18. Vasudevan S, Tong Y, Steitz JA: Switching from repression to activation: microRNAs can up-regulate translation. Science 2007, 318(5858):1931-1934.
19. Lewis BP, Burge CB, Bartel DP: Conserved seed pairing, often flanked by adenosines, indicates that thousands of human genes are microRNA targets. Cell 2005, 120(1):15-20.
20. Ambros V: The functions of animal microRNAs. Nature 2004, 431(7006):350-355.
21. Boehm M, Slack FJ: MicroRNA control of lifespan and metabolism. Cell cycle (Georgetown, Tex 2006, 5(8):837-840.
22. Calin GA, Croce CM: MicroRNA signatures in human cancers. Nature reviews cancer 2006, 6(11):857-866.
23. Jovanovic M, Hengartner MO: miRNAs and apoptosis: RNAs to die for. Oncogene 2006, 25(46):6176-6187.
24. Li Y, Wang F, Lee JA, Gao FB: MicroRNA-9a ensures the precise specification of sensory organ precursors in Drosophila. Genes & development 2006, 20(20):2793-2805.
25. Rhoades MW, Reinhart BJ, Lim LP, Burge CB, Bartel B, Bartel DP: Prediction of plant microRNA targets. Cell 2002, 110(4):513-520.
26. Lai EC: Micro RNAs are complementary to 3' UTR sequence motifs that mediate negative post-transcriptional regulation. Nature genetics 2002, 30(4):363-364.
27. Lewis BP, Shih IH, Jones-Rhoades MW, Bartel DP, Burge CB: Prediction of mammalian microRNA targets. Cell 2003, 115(7):787-798.
28. Stark A, Brennecke J, Russell RB, Cohen SM: Identification of Drosophila MicroRNA targets. PLoS biology 2003, 1(3):E60.
29. Doench JG, Sharp PA: Specificity of microRNA target selection in translational repression. Genes & development 2004, 18(5):504-511.
30. Hammell M, Long D, Zhang L, Lee A, Carmack CS, Han M, Ding Y, Ambros V: mirWIP: microRNA target prediction based on microRNA-containing ribonucleoprotein-enriched transcripts. Nature methods 2008.
31. Lim LP, Lau NC, Weinstein EG, Abdelhakim A, Yekta S, Rhoades MW, Burge CB, Bartel DP: The microRNAs of Caenorhabditis elegans. Genes & development 2003, 17(8):991-1008.
32. Xie X, Lu J, Kulbokas EJ, Golub TR, Mootha V, Lindblad-Toh K, Lander ES, Kellis M: Systematic discovery of regulatory motifs in human promoters and 3' UTRs by comparison of several mammals. Nature 2005, 434(7031):338-345.

33. Vella MC, Choi EY, Lin SY, Reinert K, Slack FJ: The C. elegans microRNA let-7 binds to imperfect let-7 complementary sites from the lin-41 3'UTR. Genes & development 2004, 18(2):132-137.
34. Chan CS, Elemento O, Tavazoie S: Revealing posttranscriptional regulatory elements through network-level conservation. PLoS computational biology 2005, 1(7):e69.
35. Kertesz M, Iovino N, Unnerstall U, Gaul U, Segal E: The role of site accessibility in microRNA target recognition. Nature genetics 2007, 39(10):1278-1284.
36. Long D, Lee R, Williams P, Chan CY, Ambros V, Ding Y: Potent effect of target structure on microRNA function. Nature structural & molecular biology 2007, 14(4):287-294.
37. Robins H, Li Y, Padgett RW: Incorporating structure to predict microRNA targets. Proceedings of the National Academy of Sciences of the United States of America 2005, 102(11):4006-4009.
38. Vella MC, Reinert K, Slack FJ: Architecture of a validated microRNA::target interaction. Chemistry & biology 2004, 11(12):1619-1623.
39. Grimson A, Farh KK, Johnston WK, Garrett-Engele P, Lim LP, Bartel DP: MicroRNA targeting specificity in mammals: determinants beyond seed pairing. Molecular cell 2007, 27(1):91-105.
40. Krek A, Grun D, Poy MN, Wolf R, Rosenberg L, Epstein EJ, MacMenamin P, da Piedade I, Gunsalus KC, Stoffel M et al: Combinatorial microRNA target predictions. Nature genetics 2005, 37(5):495-500.
41. Kiriakidou M, Nelson PT, Kouranov A, Fitziev P, Bouyioukos C, Mourelatos Z, Hatzigeorgiou A: A combined computational-experimental approach predicts human microRNA targets. Genes & development 2004, 18(10):1165-1178.
42. Rehmsmeier M, Steffen P, Hochsmann M, Giegerich R: Fast and effective prediction of microRNA/target duplexes. RNA 2004, 10(10):1507-1517.
43. Enright AJ, John B, Gaul U, Tuschl T, Sander C, Marks DS: MicroRNA targets in Drosophila. Genome biology 2003, 5(1):R1.
44. Wang X, El Naqa IM: Prediction of both conserved and nonconserved microRNA targets in animals. Bioinformatics (Oxford, England) 2008, 24(3):325-332.
45. Xu Y, Zhou X, Zhang W: MicroRNA prediction with a novel ranking algorithm based on random walks. Bioinformatics (Oxford, England) 2008, 24(13):i50-58.
46. Rajewsky N: microRNA target predictions in animals. Nature genetics 2006, 38 Suppl:S8-13.
47. Sethupathy P, Megraw M, Hatzigeorgiou AG: A guide through present computational approaches for the identification of mammalian microRNA targets. Nature methods 2006, 3(11):881-886.
48. Hendrickson DG, Hogan DJ, Herschlag D, Ferrell JE, Brown PO: Systematic identification of mRNAs recruited to argonaute 2 by specific microRNAs and corresponding changes in transcript abundance. PLoS ONE 2008, 3(5):e2126.
49. Ikeda K, Satoh M, Pauley KM, Fritzler MJ, Reeves WH, Chan EK: Detection of the argonaute protein Ago2 and microRNAs in the RNA induced silencing complex (RISC) using a monoclonal antibody. Journal of immunological methods 2006, 317(1-2):38-44.
50. Karginov FV, Conaco C, Xuan Z, Schmidt BH, Parker JS, Mandel G, Hannon GJ: A biochemical approach to identifying microRNA targets. Proceedings of the National Academy of Sciences of the United States of America 2007, 104(49):19291-19296.
51. Baek D, Villen J, Shin C, Camargo FD, Gygi SP, Bartel DP: The impact of microRNAs on protein output. Nature 2008, 455(7209):64-71.
52. Selbach M, Schwanhausser B, Thierfelder N, Fang Z, Khanin R, Rajewsky N: Widespread changes in protein synthesis induced by microRNAs. Nature 2008, 455(7209):58-63.

53. Ruike Y, Ichimura A, Tsuchiya S, Shimizu K, Kunimoto R, Okuno Y, Tsujimoto G: Global correlation analysis for micro-RNA and mRNA expression profiles in human cell lines. Journal of human genetics 2008, 53(6):515-523.
54. Matys V, Fricke E, Geffers R, Gossling E, Haubrock M, Hehl R, Hornischer K, Karas D, Kel AE, Kel-Margoulis OV et al: TRANSFAC: transcriptional regulation, from patterns to profiles. Nucleic acids research 2003, 31(1):374-378.
55. Bryne JC, Valen E, Tang MH, Marstrand T, Winther O, da Piedade I, Krogh A, Lenhard B, Sandelin A: JASPAR, the open access database of transcription factor-binding profiles: new content and tools in the 2008 update. Nucleic acids research 2008, 36(Database issue):D102-106.
56. Yu X, Lin J, Zack DJ, Mendell JT, Qian J: Analysis of regulatory network topology reveals functionally distinct classes of microRNAs. Nucleic acids research 2008, 36(20):6494-6503.
57. Shalgi R, Lieber D, Oren M, Pilpel Y: Global and local architecture of the mammalian microRNA-transcription factor regulatory network. PLoS computational biology 2007, 3(7):e131.
58. Cui Q, Yu Z, Pan Y, Purisima EO, Wang E: MicroRNAs preferentially target the genes with high transcriptional regulation complexity. Biochemical and biophysical research communications 2007, 352(3):733-738.
59. Marson A, Levine SS, Cole MF, Frampton GM, Brambrink T, Johnstone S, Guenther MG, Johnston WK, Wernig M, Newman J et al: Connecting microRNA genes to the core transcriptional regulatory circuitry of embryonic stem cells. Cell 2008, 134(3):521-533.
60. Ozsolak F, Poling LL, Wang Z, Liu H, Liu XS, Roeder RG, Zhang X, Song JS, Fisher DE: Chromatin structure analyses identify miRNA promoters. Genes & development 2008, 22(22):3172-3183.
61. Bulyk ML: Computational prediction of transcription-factor binding site locations. Genome biology 2003, 5(1):201.
62. Wasserman WW, Sandelin A: Applied bioinformatics for the identification of regulatory elements. Nature Reviews Genetics 2004, 5(4):276-287.
63. Venter JC, Adams MD, Myers EW, Li PW, Mural RJ, Sutton GG, Smith HO, Yandell M, Evans CA, Holt RA et al: The sequence of the human genome. Science 2001, 291(5507):1304-1351.
64. Barabasi AL, Oltvai ZN: Network biology: understanding the cell's functional organization. Nat Rev Genet 2004, 5(2):101-113.
65. Ravasz E, Somera AL, Mongru DA, Oltvai ZN, Barabasi AL: Hierarchical organization of modularity in metabolic networks. Science 2002, 297(5586):1551-1555.
66. Segal E, Shapira M, Regev A, Pe'er D, Botstein D, Koller D, Friedman N: Module networks: identifying regulatory modules and their condition-specific regulators from gene expression data. Nature genetics 2003, 34(2):166-176.
67. Shen-Orr SS, Milo R, Mangan S, Alon U: Network motifs in the transcriptional regulation network of Escherichia coli. Nature genetics 2002, 31(1):64-68.
68. Stark A, Brennecke J, Bushati N, Russell RB, Cohen SM: Animal MicroRNAs confer robustness to gene expression and have a significant impact on 3'UTR evolution. Cell 2005, 123(6):1133-1146.
69. Zhou Y, Ferguson J, Chang JT, Kluger Y: Inter- and intra-combinatorial regulation by transcription factors and microRNAs. BMC genomics 2007, 8:396.
70. Miska EA, Alvarez-Saavedra E, Abbott AL, Lau NC, Hellman AB, McGonagle SM, Bartel DP, Ambros VR, Horvitz HR: Most Caenorhabditis elegans microRNAs are individually not essential for development or viability. PLoS genetics 2007, 3(12):e215.

71. Nielsen CB, Shomron N, Sandberg R, Hornstein E, Kitzman J, Burge CB: Determinants of targeting by endogenous and exogenous microRNAs and siRNAs. RNA 2007, 13(11):1894-1910.
72. Cui Q, Yu Z, Purisima EO, Wang E: Principles of microRNA regulation of a human cellular signaling network. Molecular systems biology 2006, 2:46.
73. Hornstein E, Shomron N: Canalization of development by microRNAs. Nature genetics 2006, 38 Suppl:S20-24.
74. Garcia P, Frampton J: Hematopoietic lineage commitment: miRNAs add specificity to a widely expressed transcription factor. Developmental cell 2008, 14(6):815-816.
75. Milo R, Shen-Orr S, Itzkovitz S, Kashtan N, Chklovskii D, Alon U: Network motifs: simple building blocks of complex networks. Science 2002, 298(5594):824-827.
76. Yeger-Lotem E, Sattath S, Kashtan N, Itzkovitz S, Milo R, Pinter RY, Alon U, Margalit H: Network motifs in integrated cellular networks of transcription-regulation and protein-protein interaction. Proceedings of the National Academy of Sciences of the United States of America 2004, 101(16):5934-5939.
77. Johnston RJ, Jr., Chang S, Etchberger JF, Ortiz CO, Hobert O: MicroRNAs acting in a double-negative feedback loop to control a neuronal cell fate decision. Proceedings of the National Academy of Sciences of the United States of America 2005, 102(35):12449-12454.
78. Wu J, Xie X: Comparative sequence analysis reveals an intricate network among REST, CREB and miRNA in mediating neuronal gene expression. Genome biology 2006, 7(9):R85.
79. Liang H, Li WH. MicroRNA regulation of human protein protein interaction network. RNA. 2007 Sep;13(9):1402-8. Epub 2007 Jul 24.
80. Watts DJ, Strogatz SH: Collective dynamics of 'small-world' networks. Nature 1998, 393(6684):440-442.
81. Hsu CW, Juan HF, Huang HC: Characterization of microRNA-regulated protein-protein interaction network. Proteomics 2008, 8(10):1975-1979.
82. Sinha AU, Kaimal V, Chen J, Jegga AG: Dissecting microregulation of a master regulatory network. BMC genomics 2008, 9:88.

CHAPTER 9

ANALYSIS OF MICRORNA PROFILING DATA WITH SYSTEMS BIOLOGY TOOLS

Yuriy Gusev[1], Tom Schimdgen[2], Megan Lerner[1] and Dan Brackett[1]

[1]*University of Oklahoma Health Sciences Center*
Oklahoma City, Oklahoma
[2]*Ohio State University, Columbus, Ohio*

In this chapter we discuss current approaches to the analysis and biological interpretation of microRNA profiling data utilizing systems biology tools. Using Gene Ontology and Pathway Enrichment analysis in combination with the knowledge mining of cancer-related information we have developed an analysis pipeline that allows for global functional profiling of differentially expressed microRNA in cancer. To illustrate the utility of these methods, we review the main results of our recently published studies including comparison of predicted targets for five published datasets of aberrantly expressed microRNAs in human cancers and in-depth analysis of functions and pathways affected by an overexpressed cluster of miRNAs in hepatocellular carcinoma. Using combinatorial enrichment analysis – a modification of GO enrichment analysis, we have identified Gene Ontology categories as well as biological functions, disease categories, toxicological categories, and signaling pathways that are: targeted by multiple microRNAs; statistically significantly enriched with target genes; and known to be affected in specific cancers. Our results suggest that co-expressed miRNAs provide systemic compensatory response to the cancer-specific phenotypic changes by downregulating functional categories and signaling pathways that are known to be abnormally activated in a particular cancer.

1. Introduction

microRNA (miRNA) has recently emerged as new and important class of cellular regulators. Experimental studies have provided strong evidence that aberrant expression of miRNA is associated with a broad spectrum of human diseases, including cancer,[1] diabetes,[2] cardiovascular,[3] and psychological disorders.[4]

The small number of miRNAs discovered in human (~800 miRNAs, miRBase 12.0)[5] hold a big potential as a new type of biomarkers because of a relative abundance of this class of molecules, high specificity of tissue expression, and involvement in regulation of a large number of human genes (up to 80% of known genes by the most recent estimates[6]).

New technologies have been developed recently that allow measurement of the level of expression for all known miRNAs in relatively small RNA samples providing the opportunity for global miRNA profiling in normal and cancer cells as well as in the clinical tumor specimens (reviewed in chapters 1-4). Global profiling of miRNA expression has been shown to provide a more accurate method of classifying cancer subtypes than expression profiling of all protein-coding genes.

Further characterization of these abnormal miRNA expression signatures and identification of the most significant and informative aberrantly expressed miRNAs has most recently stimulated commercial development of new tissue- and biofluid-specific diagnostic markers and a new type of miRNA-based drugs .

However, once the microRNA profiling data are generated, it creates a downstream problem of biological interpretation of function for not only individual miRNAs but also for the large groups of differentially expressed miRNAs. This problem is very challenging because little is known about the specific functions of microRNAs, the predicted number of targets is very large (on average ~200 targets[6] per miRNA) and only limited number of targets has been characterized experimentally.

2. Computational Framework for Functional Analysis of microRNA Targets

Several high throughput technologies such as microarrays, real-time PCR and deep sequencing allow screening of tumor samples for expression of multiple or all known miRNAs (see chapters 1-4). Independently of the technological platform and the statistical analysis method used, the end point of any global expression profiling experiment is a list of differentially expressed miRNAs. Then, a target prediction algorithm of choice can be used to generate a list of potential target genes (reviewed in[7]).

Briefly, for each miRNA a list of potential target sites (with evolutionary conservation requirement as an additional constrain) is determined on the 3'-UTRs of human gene-coding transcripts based on a search for complimentary binding sites for the sequence of "seed" regions of the 5'ends of mature miRNA (positioned at nucleotides 2-8). Some additional criteria for target prediction may vary depending on algorithm and usually includes thermodynamics-based constrains for RNA-RNA interactions.[8]

Computational prediction of targets for individual miRNAs or for a miRNA group (cluster) of interest can be performed using stand alone open access program (TargetScan,[6] DIANA microT,[9] PicTar,[10] miRanda[11]), or publicly available software suites allowing to combine predictions of several popular prediction algorithms (MAMI,[12] MirGen[13]). Until recently many prediction algorithms have provided a relatively high rate of false positive predictions. However the newly developed methods of experimental target validation (SILAC,[14] AIN-IP[15]) have recently generated a significant number of newly identified targets allowing for further refinement of existing predictions and development of new more accurate algorithms such as mirWIP[15] When applied to the analysis of miRNA profiling data, even the best of these prediction programs generate large lists of predicted targets of each individual miRNA.

The number of experimentally validated targets is also growing rapidly. The database of experimentally supported targets TarBase[16] has listed 1100 human genes that are validated miRNA targets.

Prediction algorithms and experimental validation studies demonstrated that each miRNA targets multiple transcripts and many transcripts are targeted by multiple miRNAs. An average number of targets per miRNA is estimated to be approximately 200.[6] Less is known about the distribution of the number of multiple target sites within the same transcript however some of the algorithms predict that transcripts may have target sites for up to 400 different miRNAs on a single 3'UTR (unpublished data).

To illustrate the complexity of the miRNA-mediated regulatory networks we have generated a network of interactions of experimentally validated targets for a small group of 4 miRNA (miR-18a, miR-19a, miR-19b-1 and miR-20a) from the well known cistron miR-17-92. These miRNAs are all encoded by a single primary transcript and were reported to be simultaneously overexpressed in several human cancers[17] (Figure 1). This network was generated by MetaCore software (GeneGO Inc.). The miRNAs are shown on the sides of the diagram. The known experimentally validated targets are shown with arrows indicating interactions.

This diagram shows both types of relationship between the miRNAs and their targets: several miRNAs are targeting multiple targets each and several genes are targeted by two microRNAs each. It is also evident that, in spite of the fact that those 4 microRNAs are originating from the same cluster miR-17-92, their target sets are not completely overlapped.

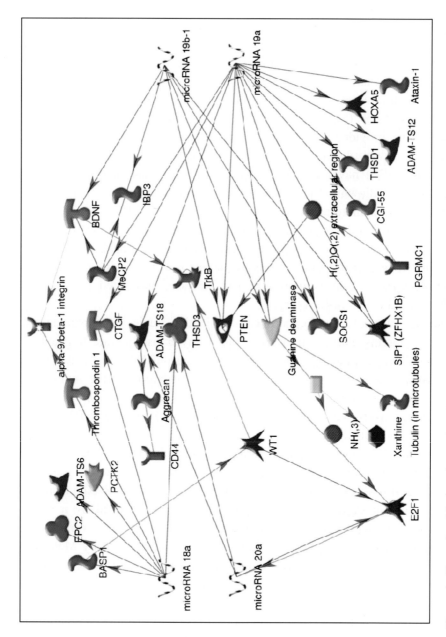

Figure 1. Network of known interactions between 4 microRNAs and their validated targets. Generated with GeneGo MetaCore (GeneGo, Inc)

This target multiplicity and many-to-many relation present a serious computational challenge that does not depend on the type of prediction algorithm and requires specialized computational methods and tools that can provide functional profiling of differentially expressed miRNAs in a high throughput fashion.

2.1. *Gene ontology enrichment analysis of microRNA targets*

Several computational methods allow functional annotation of gene lists according to functional annotation systems, such as Gene Ontology (Gene Ontology Consorcium[18]) or various databases of known signaling pathways (KEGG,[19] GeneMAPP[20]).

In order to exclude those categories that were found simply by chance, a statistical procedure known as enrichment (over-representation) analysis has been developed independently by several groups (reviewed in Draghici *et al*.[21]). Many software tools use the Fisher Exact test to estimate significance of enrichment (over-representation) in annotation terms such as Gene Ontology terms or canonical pathways.[22] For each annotation term the Fisher Exact probability is calculated using the Gaussian Hypergeometric probability distribution that describes sampling without replacement from a finite population consisting of two types of elements.

In case of global functional annotation of all known microRNA targets, a calculation of Fisher Exact p-value for each GO term is based upon a comparison of target genes that belong to particular GO term with the reference set of all predicted human target genes (annotated to any GO terms). Importantly, if the miRNA profiling data were obtain using specific microarray or other high throughput platform such as real-time quantitative PCR, than the reference set should include the union of predicted targets of only those microRNAs which probes were present on the arrays.

For example, let us consider a set of 20 miRNAs that were reported as over-expressed in colon cancer.[23] Using open access software tool MAMI (Meta prediction of microRNA targets) we generated a union of predictions of the 5 most popular target prediction algorithms. As a result we obtained a list of all target genes of these 20 microRNAs with total of 5170 genes. A functional annotation and over-representation analysis of GO Biological Process (GO BP) terms for this gene list was performed using an open access program DAVID.[24] As a reference set we used all human miRNA targets predicted by the same 4 algorithms (total of 10574 genes). Note that if the miRNA profiling data were

Table 1. Example: 2x2 Contingency table for microRNA targets counts from GO term apoptosis.

	All target genes of co-expressed microRNAs	All Human microRNA targets (or targets of all miRNAs that were included on microarray)
In GO Term	197	472
Not In GO Term	3494	10102

obtain with a specific microarray, than the reference set should include targets of only those microRNAs that were present on the microarray. For this example Table 1 represent a contingency table for GO term Apoptosis. Using DAVID software the Fisher Exact p-value was calculated (p = 0.01235) and since p-value is less than 0.05, we assume that this target list is statistically significantly enriched or overrepresented in Apoptosis GO BP term as compared to a random chance.

For this example with colon cancer the resulted list of enriched GO BP terms contains 176 categories that are significantly enriched in target genes of 20 over-expressed miRNAs.

Evidently, a number of enriched functional categories is rather large and interpretation of these results could be a challenge. In order to narrow down the list of GO terms, these categories could be ranked according to some quantitative criteria, for instance a fraction of genes targeted in a particular functional group, or a significance of over-presentation (p-values from the Fisher Exact test). However, in many cases the most general GO terms have a tendency to include a large number of target genes and to be highly significant. These general GO terms can often be found among the top ranked enriched categories. For instance, for colon cancer, a list of the top 20 most significantly enriched categories consists of mostly very general GO terms (GO BP_1 and GO BP_2) that are related to the regulation of cell processes. These broad categories normally include very large number of genes and provide only very general idea about their function.

To avoid this problem, the analysis could be restricted to a lower level of GO hierarchy providing smaller and more specific gene lists. In our example with colon cancer we have performed additional analysis restricted to only the lowest level (level 5: GO BP_5) of GO terms. The resulting list of enriched categories shows 36 more specific functional categories that are significantly over represented among targets of 20 miRNAs (Table 2). Many of these categories, such as "regulation of progression through cell cycle" or "stress-activated protein kinase signaling pathway", are well known to be affected in many cancers.

Table 2. Gene Ontology BP_5 terms that are enriched with targets of miRNAs overexpressed in Colon Cancer, standard enrichment analysis.

GO BP_5 Term	Count	P-Value
regulation of nucleobase, nucleoside, nucleotide and nucleic acid metabolism	772	2.92E-14
transcription	791	1.36E-13
biopolymer modification	629	1.22E-10
phosphate metabolism	336	4.51E-08
regulation of progression through cell cycle	195	4.46E-06
enzyme linked receptor protein signaling pathway	88	1.71E-04
negative regulation of cellular metabolism	88	2.20E-04
protein kinase cascade	121	2.53E-04
cell growth	69	9.71E-04
regulation of cell size	69	9.71E-04
regulation of protein kinase activity	63	0.0012
small GTPase mediated signal transduction	112	0.001303
negative regulation of progression through cell cycle	68	0.001726
intracellular receptor-mediated signaling pathway	26	0.002467
stress-activated protein kinase signaling pathway	26	0.003842
positive regulation of cellular metabolism	67	0.004638
chromosome organization and biogenesis	103	0.006869
positive regulation of programmed cell death	68	0.007934
vesicle-mediated transport	142	0.015351
regulation of cell motility	17	0.015427
regulation of cell migration	17	0.015427
vacuole organization and biogenesis	9	0.017697
endosome transport	14	0.017848
Notch signaling pathway	19	0.020648
apoptosis	197	0.021507
lysosomal transport	6	0.028821
cytoskeleton organization and biogenesis	138	0.033054
cartilage condensation	7	0.033699
regulation of cell shape	17	0.035196
regulation of programmed cell death	126	0.042611
brain development	18	0.045917
protein transport	198	0.047414
lipid modification	13	0.048145
cell migration	42	0.049238
secretory pathway	74	0.049295
regulation of apoptosis	125	0.049678

2.2. Pathway enrichment analysis of microRNA targets

A similar approach can be applied to the enrichment analysis of known signaling pathways. In this case a list of target genes is compared to a list of all genes from a particular pathway while a reference gene list contains all genes from a complete set of known (canonical) signaling pathways and is compared to the list of targets of all known miRNAs (or all microRNAs that are represented on the particular microarray) that belong to all known pathways.

In our example with colon cancer, we have also performed pathway enrichment analysis using DAVID software. A collection of KEGG signaling pathways was analyzed using the Fisher exact test. A total of 19 pathways were found that were enriched with microRNA target genes at significance level 0.05. The top ranked pathway (HSA04510: FOCAL ADHESION) contained 101 genes targeted by the group of 20 miRNAs that are overexpressed in colon cancer (p-value = 2.22E-05). Several other pathways from this list were reported to be affected by cancer as well.

This type of enrichment analysis can uncover some common biological themes that are present in a set of miRNA target genes, thus providing an investigator with additional clues for the follow-up experiments. However such an approach has several shortcomings as it generates a wide spectrum of many functional categories presumably affected by miRNAs which makes it difficult to interpret or validate the results. In addition, when this method is used, all of the information related to the targets of specific microRNAs is lost and functional categories can not be traced back in order to determine which miRNAs are involved in regulation of a particular functional term. Also lost is important information on those miRNAs that collectively target the same genes or the same functional categories and therefore might have bigger regulatory impact.

2.3. A modification of GO enrichment analysis accounting for combinatorial targeting of Gene Ontology categories by multiple miRNAs

In this section we discuss modification of enrichment analysis which allows us to focus on those functional categories that are targeted by multiple microRNAs and are likely to be affected the most. This method of statistical analysis was first proposed and implemented in the target analysis program miRgate (Delfour *et al.*, 2007).

In order to determine association between GO category and miRNA, the algorithm has adopted the hypergeometric distribution. First, The GO categories

Table 3. Gene Ontology BP terms that are enriched with targets of miRNAs overexpressed in Colon Cancer and targeted by 100% of miRNAs in the group. Results of combinatorial enrichment analysis.

GO Category	GO #	Observed	Expected	p-value	Genes	#MicroRNAs
induction of apoptosis	GO:0006917	146	115.3307	0.00215	39	100% [20/20]
cell division	GO:0051301	172	143.407	0.00789	53	100% [20/20]
homophilic cell adhesion	GO:0007156	150	126.327	0.0161	30	100% [20/20]
positive regulation of transcription, DNA-dependent	GO:0045893	135	116.0139	0.0351	41	100% [20/20]
ubiquitin cycle	GO:0006512	430	380.6077	0.00506	127	100% [20/20]
intracellular protein transport	GO:0006886	199	176.9164	0.0437	65	100% [20/20]
metabolic process	GO:0008152	300	269.5714	0.0288	120	100% [20/20]
cell cycle	GO:0007049	364	329.1725	0.0247	131	100% [20/20]

are determined for predicted targets of each miRNAs from a group. But the significance of overrepresentation for each category is calculated based on a different criteria: a number of "hits" by multiple microRNAs, thus directly utilizing information about how many miRNAs are targeting the same category. Using a selected threshold for p-values of hypergeometric distribution this technique also allows for filtering out those functional categories that were found simply by chance. The list of overrepresented categories could be additionally trimmed by selecting only those overrepresented GO categories which are targeted by at least a given number or fraction of co-expressed miRNAs. This allows for obtaining a short list of functional categories that are more likely to be affected by a given group of co-expressed miRNAs.

A resulting list of target genes from overrepresented GO categories can be subjected to a more detailed functional analysis such as pathway analysis using a wide spectrum of publicly available or commercial software tools to obtain a more refined biological interpretation of affected gene categories. Importantly, by retaining information on specific miRNAs that target a particular GO category, one can selectively restrict analysis by only those functional categories that are targeted by a subset of co-expressed miRNAs or by all miRNAs in a cluster.

As an example, we have applied this method to the analysis of the same set of 20 miRNAs that are overexpressed in colon cancer. We have determined enriched Gene Ontology categories that are targeted by multiple miRNAs in the group. Assuming that the categories that are targeted by the all miRNAs in the group are affected the most we have selected only those overrepresented GO terms that are targeted by all 20 miRNAs. The resulting short list of GO BP terms has contained only 8 categories (Table 3), most of which are well-known cancer related biological processes such as induction of apoptosis (top ranked category), cell division and cell cycle.

We have applied this method of combinatorial enrichment analysis to the analysis of several miRNA profiling datasets generated by our group[25] or published by others.[23]

3. Applications of Combinatorial Enrichment Method for the Analysis of miRNA Profiling Data in Human Cancers

Multiple studies have reported the results of functional annotation of genes targeted by single miRNAs,[26] all known miRNAs,[27,28] or small clusters of miRNAs that were selected based on high similarity of "seed" sequences in the 5' region and/or large overlap of predicted target sets.[29] However, in case of experimentally obtained miRNA profiling data these approaches are not very practical when the task is to determine common biological functions and regulatory pathways that are targeted by experimentally detected groups of co-expressed miRNAs. Specifically in cancer such groups of miRNAs are often much larger, have fewer common target genes, and do not share similar "seed" sequences.

In our recent studies[30,31] we addressed this problem of biological interpretation of miRNA profiling data using enrichment analysis of biological processes, disease categories and signaling pathways that are targeted collectively by co-expressed miRNAs. Here we present a summary of the results from these studies.

Five groups of co-expressed miRNAs were selected from the literature for this study: 3 groups that were reported[23] as being over-expressed in breast cancer (14 miRNAs), colon cancer (20 miRNAs) and lung cancer (33 miRNAs). We have also included a set of miRNAs that we found to be significantly over-expressed in pancreatic cancer (47 miRNAs)[25]. An additional group of seven miRNAs was reported as being over-expressed in lymphomas.[32] This group of miRNAs is encoded by a single gene (cistron miR-17- 92) and expressed as a single primary transcript. Over-expression of cistron miR-17-92 was found in B-lymphomas and also was shown to have strong correlation with T-lymphoma development in an animal model [33]. To avoid possible bias of sample size, these datasets were selected to represent the whole spectrum of group sizes and seed similarity of co-expressed miRNAs that are observed in cancers: from a small family of co-expressed miRNAs (7 miRNAs, cistron miR-17-92) to a large and diverse group of co-expressed miRNAs (47 miRNAs, pancreatic cancer).

3.1. Combinatorial enrichment analysis of microRNA targets n human cancers

The Gene Ontology (GO) enrichment analysis of biological processes targeted by each of five groups of miRNAs was performed using combinatorial enrichment analysis that is specifically designed to take in account information about the number of miRNAs that are targeting the same GO categories i.e. number of miRNA "hits" per GO category (Delfour et al., 2007). The GO categories were determined for predicted targets of each miRNAs from a group. This set of GO categories was filtered based upon significance of overrepresentation of "hits' by multiple microRNAs using a selected threshold for p-values of hypergeometric distribution ($p \leq 0.05$). All significant categories were then compared based between five data sets. The sets of enriched categories for each cancer had surprisingly low overlap with other cancers demonstrating specificity of collective effect of co-expressed miRNA for each cancer (Fig. 2).

An additional filter was applied to select only those overrepresented GO categories which are targeted by 100% of miRNAs in a group. The resulting list of the enriched GO terms categories for 4 cancers is presented in Table 4.

For all data sets, the top ranked GO terms that are targeted by 100% of miRNA include different subsets of biological processes related to cell proliferation that are commonly known to be associated with various types of cancer, including apoptosis, cell cycle, cell proliferation. Other categories were more specific for each cancer such as regulation of Rho protein signal transduction for pancreatic cancer.[34]

3.2. Systems biology in depth analysis and comparison of predicted miRNA targets with ingenuity IPA

To further evaluate the specific functional profiles of genes from the broad GO categories that are targeted by miRNAs, we performed more detailed functional analysis using Ingenuity Pathway Analysis system (IPA 5.5, Ingenuity Systems, Redwood, CA).

For each set of co-expressed miRNAs we have generated 3 sets of predicted gene targets using combinatorial enrichment analysis as a statistical filter to reduce number of gene and limit it to only those enriched GO categories that are

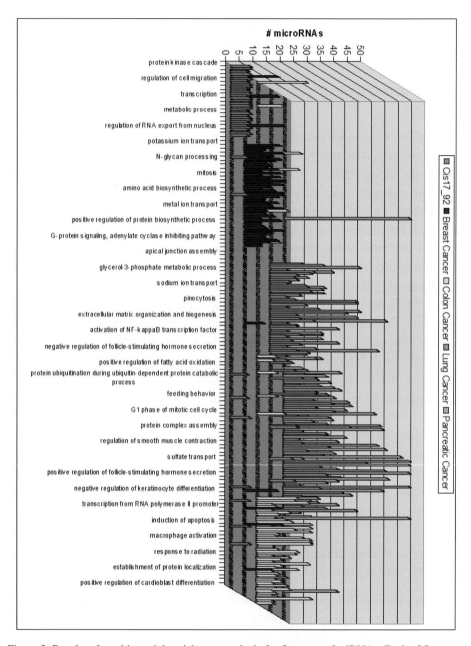

Figure 2. Results of combinatorial enrichment analysis for 5 groups of miRNAs. Each of 5 groups of data is color-coded according to the type of cancer.
X-axis – GO categories that are enriched with targets of multiple miRNAs
Y-axis – Number of miRNAs targeting particular GO category.

Table 4. Results of combinatorial enrichment analysis of microRNA targets for 4 human cancers. Gene ontology categories significantly enriched with targets of multiple miRNAs.

GO Category	p-value	#genes	#microRNAs
PANCREATIC CANCER			
transport	0.00038	247	100% [47/47]
regulation of Rho protein signal transduction	0.00133	36	100% [47/47]
small GTPase mediated signal transduction	0.0107	138	100% [47/47]
protein complex assembly	0.0189	61	100% [47/47]
actin cytoskeleton organization and biogenesis	0.0407	54	100% [47/47]
proteolysis	0.0461	171	100% [47/47]
positive regulation of I-kappaB kinase/NF-kappaB cascade	0.0488	46	100% [47/47]
LUNG CANCER			
protein targeting	0.00799	35	94% [31/33]
regulation of cyclin-dependent protein kinase activity	0.0321	19	91% [30/33]
sodium ion transport	0.036	46	100% [33/33]
protein transport	0.0434	157	100% [33/33]
cell proliferation	0.0444	142	100% [33/33]
apoptosis	0.0488	150	97% [32/33]
COLON CANCER			
induction of apoptosis	0.00215	39	100% [20/20]
ubiquitin cycle	0.00506	127	100% [20/20]
cell division	0.00789	53	100% [20/20]
homophilic cell adhesion	0.0161	30	100% [20/20]
cell cycle	0.0247	131	100% [20/20]
metabolic process	0.0288	120	100% [20/20]
positive regulation of transcription, DNA-dependent	0.0351	41	100% [20/20]
intracellular protein transport	0.0437	65	100% [20/20]
BREAST CANCER			
transforming growth factor beta receptor signaling pathway	9.99E-04	13	100% [14/14]
inflammatory response	2.12E-03	51	100% [14/14]
small GTPase mediated signal transduction	5.88E-03	102	100% [14/14]

Table 4 (*Continued*)

GO Category	p-value	#genes	#microRNAs
glycogen metabolic process	8.11E-03	7	100% [14/14]
ubiquitin cycle	9.78E-03	138	100% [14/14]
negative regulation of transcription from RNA polymerase II promoter	1.35E-02	60	100% [14/14]
regulation of translation	2.07E-02	16	100% [14/14]
mitosis	2.41E-02	37	100% [14/14]
DNA recombination	2.77E-02	11	100% [14/14]
unfolded protein response	3.40E-02	7	100% [14/14]
cell cycle arrest	3.87E-02	25	100% [14/14]
nervous system development	4.28E-02	123	100% [14/14]
potassium ion transport	4.34E-02	49	100% [14/14]
cell division	4.50E-02	48	100% [14/14

collectively targeted by co-expressed miRNAs:

i. Targets that belong to GO categories targeted by at least one of the co-expressed miRNAs;
ii. Targets that belong to GO categories targeted by at least 50% of the co-expressed miRNAs;
iii. Targets that belong to GO categories targeted by 100% of the co-expressed miRNAs.

We started with raw lists of the unions of all targets predicted for each of the miRNAs in a range from 2175 genes (cistron mir-17-92) to 5356 genes (pancreatic cancer). Filtering raw sets of predicted targets by combinatorial enrichment analysis provided a significant reduction of target lists in a range from 2.5 fold to over 4 fold. Three sets of genes were generated for each of 5 groups of miRNAs and were then analyzed by Ingenuity Pathway Analysis tools to determine more detailed information about biological functions, disease categories, toxicological categories, canonical signaling pathways, and drugs associated with each data set.

3.2.1. *Reference set of genes known to be affected in cancer*

To select the most important functional categories and pathways for analysis and to understand relevance of miRNA-targeted categories to the specific cancer, we generated reference sets of genes for each of five cancers by keyword search of the Ingenuity Knowledge Base. This search provided us with very conservative lists of genes that were reported to be affected in a specific cancer by multiple research groups and were then manually curated by a group of expert biologists. These 5 reference sets of genes known to be affected in lymphoma, breast cancer, colon cancer, lung cancer, and pancreatic cancer were used in the Ingenuity Pathway Analysis system to generate sets of overrepresented biological functions, disease categories, and pathways, that are known to be affected in each of these cancers. These cancer specific categories were compared with the categories and pathways predicted to be affected by co-expressed miRNAs.

3.2.2. *Comparative analysis of biological functions and disease categories*

Using Ingenuity Pathway Analysis system (IPA 5.0) we have compared target gene sets determined by combinatorial enrichment algorithm with the reference groups of genes that are known to be affected in a specific cancer. This allowed us to determine which top ranked categories of reference sets would be statistically enriched with miRNA targets. The results indicate that the majority of top ranked biological functions and disease categories as well as toxicological categories that were tissue specific for each specific cancer were also statistically significantly overrepresented in our target lists. Top biological functions and disease categories were compared among 5 groups of data, using gene lists from GO categories targeted by at least 50% of miRNAs in the group. The top ranked disease category for all 5 datasets was Cancer with highly significant enrichment ($p \sim E-10 \div E-20$). Within this top category, we identified a significant number of miRNA targets that are known as tissue specific biomarkers of each of the five cancers as well as a large number of miRNA targets that are known to be affected in other cancers.

For instance, for pancreatic cancer a list of miRNA targets included both k-ras and p53 genes (Table 5) that are well known genes involved with pancreatic.

A list of miRNA targets for colon cancer has included the APC (adenomatosis polyposis coli) gene – a well-known biomarker of colon cancer among other well known oncogenes.

Overall, in our analysis we identified 25 known cancer-related genes that have been already experimentally validated as targets of miRNAs.

The top ranked biological functions included general categories of Cell Cycle, Cell Death, Cell Morphology, as well as more specific Post-Translational Modification and DNA Replication, Recombination and Repair. The detailed results and statistics of IPA enrichment analysis of functional categories for all datasets are published by authors elsewhere.[30]

Table 5. List of genes from enriched GO BP categories that are targeted by at least 50% of overexpressed miRNAs and known to be affected in Pancreatic Cancer.

Target Name	Description	Location	Type	Drugs
AKT1	v-akt murine thymoma viral oncogene homolog 1	Cytoplasm	kinase	enzastaurin
BCL2	B-cell CLL/lymphoma 2	Cytoplasm	other	oblimersen
CCKBR	cholecystokinin B receptor	Plasma Membrane	G-protein coupled receptor	CR 2945
CDK6	cyclin-dependent kinase 6	Nucleus	kinase	flavopiridol
CTSB	cathepsin B	Cytoplasm	peptidase	
CTSL2	cathepsin L2	Cytoplasm	peptidase	
E2F1	E2F transcription factor 1	Nucleus	transcription regulator	
FAS	Fas (TNF receptor superfamily, member 6)	Plasma Membrane	transmembrane receptor	
HGF	hepatocyte growth factor (hepapoietin A; scatter factor)	Extracellular Space	growth factor	
HMGA1	high mobility group AT-hook 1	Nucleus	transcription regulator	
IL6	interleukin 6 (interferon, beta 2)	Extracellular Space	cytokine	
KRAS	v-Ki-ras2 Kirsten rat sarcoma viral oncogene homolog	Cytoplasm	enzyme	
PLAU	plasminogen activator, urokinase	Extracellular Space	peptidase	
RHOB	ras homolog gene family, member B	Cytoplasm	enzyme	
THOC1	THO complex 1	Nucleus	other	
TOP1	topoisomerase (DNA) I	Nucleus	enzyme	elsamitrucin, T 0128, CT-2106, BN 80927, tafluposide, TAS-103, irinotecan, topotecan, 9-amino-20-camptothecin, rubitecan, gimatecan, karenitecin
TP53	tumor protein p53 (Li-Fraumeni syndrome)	Nucleus	transcription regulator	

3.2.3. Comparative analysis of toxicology categories

Using IPA we have also analyzed top ranked toxicology related gene lists for each of the five reference gene lists and compared them with toxicology categories found in our miRNA target lists.

We found that 8 top ranked toxicology categories for each cancer were statistically significantly overrepresented among miRNA targets. We found it particularly interesting that several categories related to oxidative stress and hypoxia were among the top ranked overrepresented categories for miRNA targeted genes. These findings are in agreements with recent experimental data reporting over expression of multiple miRNAs in response to oxidative stress or hypoxia[35] and showing a functional link between hypoxia, a well-known tumor microenvironment factor, and microRNA expression.

3.2.4. Enrichment analysis of signaling pathways

To further evaluate the specific functions of genes from the broad GO categories that are targeted by miRNAs, we performed additional, more detailed pathway analysis (IPA 5.0, Ingenuity Systems). We compared gene sets determined by GO enrichment algorithm against known signaling pathways to determine which pathways would be statistically enriched with miRNA targets. We were also interested in determining which pathways were affected the most by multiple miRNAs from the same co-expressed group in each specific cancer.

The results of the pathway analysis were similar for all datasets. We found that a large fraction of top ranked pathways known to be affected in cancer were also collectively targeted by the groups of co-expressed miRNAs.[30] However, when only the significantly enriched pathways were compared ($p \leq 0.05$), the list of pathways were more specific for each type of cancer.[30]

The detailed inspection of pathway diagrams revealed an interesting pattern of genes targeted by miRNAs. In the majority of inspected overrepresented pathways some of the miRNA targets were the same known genes that are associated with this cancer. However, many targets were not reported as associated by a cancer and were found among the genes that are directly downstream and/or upstream of the cancer related genes in the same branches of signaling cascades (Figures 3, 4).

One typical example is shown for colon cancer (Fig. 3) and one for pancreatic cancer (Fig. 4).

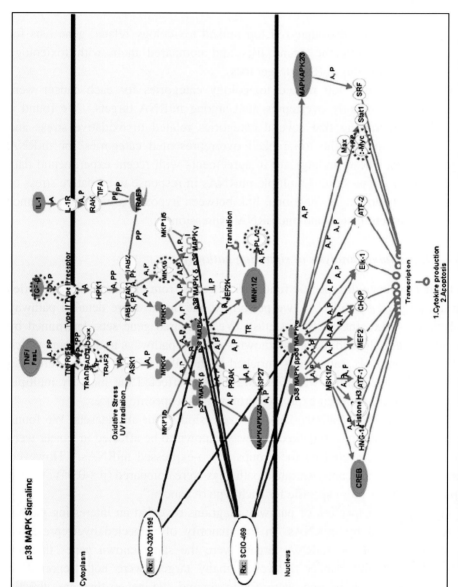

Figure 3. P38 MAPK Pathway. miRNA targets – solid green; Colon cancer associated genes – traced with red.

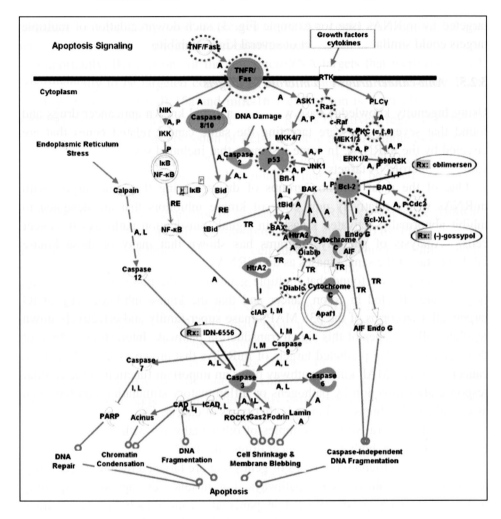

Figure 4. Apoptosis signaling pathway. miRNA targets – solid green; Pancreatic caner related genes – traced with red.

In many target-enriched pathways we found that miRNAs are targeting multiple protein kinases that are important mediators of signal transduction pathways and are often targeted by anticancer drugs known as kinase inhibitors (specific discussion is provided in the next section). Since all miRNAs in this study were over-expressed in cancer, our findings suggest that their overall effect would be to down-regulate many or, sometimes, all of the abnormally activated alternative signal transduction cascades in many of the pathways known to be affected by a particular cancer. For those pathways where multiple kinases are

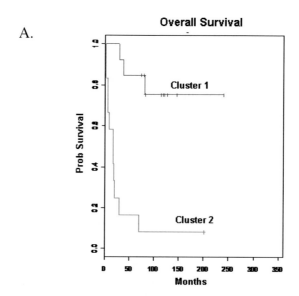
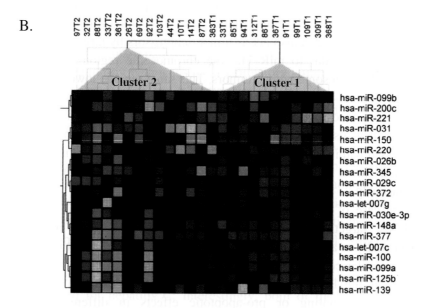

Figure 5. Prognostic value of miRNA expression in patients with HCC[43].
A. Kaplan-Meyer survival curves for two groups of patients. Red line: data for patients with higher survival time (cluster 1) Green line: patients with poorer survival; $P < 0.05$ (cluster 2).
B. Heatmap of the results of hierarchical clustering of tumor samples according to miRNA expression values. Samples from the cluster 1 were clustered together and show elevated level of expression for many miRNA.

4. Analysis of miRNAs Profiling Data in Hepatocellular Carinomas

In our recent study of microRNA expression in hepatocellular carcinomas we have analyzed 25 samples of HCC tumors.[43] Mature microRNA were profiled by a high throughput real-time quantitative PCR. miRNA profiling data was analyzed using semi supervised principal component analysis - predictive analysis of microarrays[44] (PAM). Importantly, by using PAM algorithm we have identified a cluster of 19 microRNAs with significant prognostic value. When overrexpressed in tumors these miRNAs were statistically significantly ($p = 0.0293$) associated with better overall survival of HCC patients (Fig. 5).

4.1. Analysis of functional categories targeted by cluster of 19 miRNAs associated with better survival in HCC

Combinatorial enrichment analysis of predicted targets for this cluster of "better survival signature" miRNAs has shown that only 3 Gene Ontology categories were significantly enriched ($p < 0.05$) and were targeted by at least 80% of miRNAs from the cluster (Table 6).

Table 6. Significantly enriched GO categories targeted by cluster of 19 miRNAs.

Biological Process	Target genes	Significance	MicroRNA cluster members targeting the same function
cell division [GO:0051301]	61	1.74E-02	100% [19/19]
mitosis [GO:0007067]	46	1.26E-02	100% [19/19]
G1/S transition of mitotic cell cycle [GO:0000082]	15	3.17E-02	84% [16/19]

All 3 categories were related to regulation of cell division and cell cycle progression. We identified 86 genes from these functional categories that were predicted miRNA targets. 39 genes were already reported in the literature as associated with several types of human cancers.

Eight of these genes were reported to be specifically associated with liver cancer: ACVR1B, CCND1, CDC25A, CDKN3, HGF, PLK1, RAN, and TPX2.

4.2. *Pathway analysis of predicted targets of 19 miRNAs in HCC*

Targets from these 3 GO categories were further analyzed using Ingenuity Pathway Analysis. The top ranked list of significantly enriched pathways has included G1/S signaling pathway. Four genes were identified as targets of microRNAs from comparison with reference set of HCC-associated genes has shown that in case of overexpression of miRNAs these 4 target genes will be downregulated and effectively downregulate all branches of abnormally activated signaling transduction cascades of G1/S pathway (Fig. 6).

Comparison with reference set of HCC-associated genes has shown that in case of overexpression of miRNAs these 4 target genes will be downregulated and effectively downregulate all branches of abnormally activated signaling transduction cascades of G1/S pathway (Fig. 6). Using Ingenuity knowledge bas we have identified several known anticancer drugs that are specifically targeting the same four nodes of G1/S pathways as do the miRNAs that were identified in our analysis. Experimental evidence related to these drugs show that the main effect of downregulation of expression of these nodes was indeed the inhibition of cell proliferation and tumor growth.

To summarize, the systems biology analysis of miRNA profiling in HCC has facilitated biologically relevant interpretation of the data, pinpointed a short list of genes and pathways that could be relevant to the observed difference in disease outcome, and provided testable hypothesis for the validation experiments.

5. Conclusion

In this chapter we discussed existing challenges of interpretation of miRNA profiling data and presented some of the available computational tools to address these problems. Using an array of systems biology tools that includes enrichment analysis and pathway analysis as well as knowledge mining of published information related to cancer it is feasible to filter large sets of predicted microRNA targets and to narrow down potential candidate pathways for further experimental validation.

We have also tested an idea that by focusing on functional categories and pathways that are targeted by multiple microRNAs, we might be able to determine those biological functions that could be affected the most by a group of differentially expressed miRNAs and also could be more specific for each particular cancer.

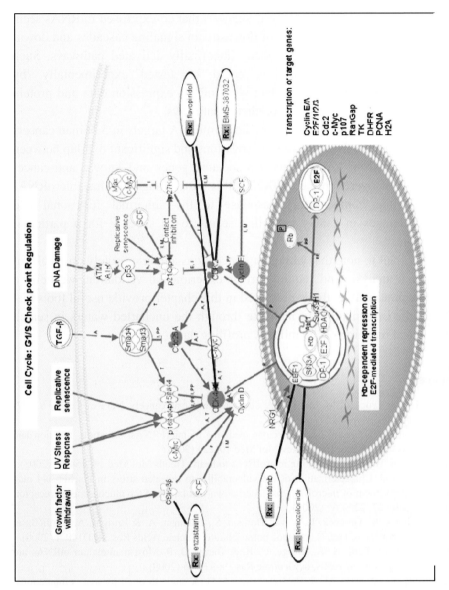

Figure 6. G1/S Check Point Pathway. Predicted miRNA targets are shown in solid green; drugs are shown as white ovals with labels; drug targets are indicated by blue arrows.

The results of our pathway analysis suggests that co-expressed miRNAs seem to collectively target a broad range of downstream signaling cascades and down-regulating expression of genes in many abnormally activated pathways. Such computationally derived hypothesis could be tested experimentally, by comparing microRNA expression data with mRNA expression data and protein expression data for the short list of predicted pathways.

Our systems biology analysis of predicted miRNA targets in 5 human cancers has revealed an interesting pattern of correlations and significant overlap between targets of microRNA, known cancer-associated genes and known anti-cancer drugs. In case of overexpressed miRNAs this pattern suggests that microRNAs provide systemic compensatory response to the abnormal functional and phenotypic changes in cancer cells that seems to be specific for a particular cancer.

It is important to note that while our conclusions are based solely on results of computational analysis and require further experimental validation, we believe that the computational methods presented in this chapter provide useful tools that could help the investigators to navigate through the uncharted waters in search for a biological meaning of the miRNA profiling data.

References

1. Calin, G. A. & Croce, C. M. MicroRNA signatures in human cancers. Nat Rev Cancer 6, 857-66 (2006).
2. Hennessy, E. & O'Driscoll, L. Molecular medicine of microRNAs: structure, function and implications for diabetes. Expert Rev Mol Med 10, e24 (2008).
3. Callis, T. E. & Wang, D. Z. Taking microRNAs to heart. Trends Mol Med 14, 254-60 (2008).
4. Uchida, S. et al. Characterization of the vulnerability to repeated stress in Fischer 344 rats: possible involvement of microRNA-mediated down-regulation of the glucocorticoid receptor. Eur J Neurosci 27, 2250-61 (2008).
5. Griffiths-Jones, S., Grocock, R. J., van Dongen, S., Bateman, A. & Enright, A. J. miRBase: microRNA sequences, targets and gene nomenclature. Nucleic Acids Res 34, D140-4 (2006).
6. Friedman, R. C., Farh, K. K., Burge, C. B. & Bartel, D. P. Most mammalian mRNAs are conserved targets of microRNAs. Genome Res 19, 92-105 (2009).
7. Sethupathy, P., Megraw, M. & Hatzigeorgiou, A. G. A guide through present computational approaches for the identification of mammalian microRNA targets. Nat Methods 3, 881-6 (2006).
8. Bartel, D. P. MicroRNAs: target recognition and regulatory functions. Cell 136, 215-33 (2009).
9. Kiriakidou, M. et al. A combined computational-experimental approach predicts human microRNA targets. Genes Dev 18, 1165-78 (2004).
10. Krek, A. et al. Combinatorial microRNA target predictions. Nat Genet 37, 495-500 (2005).
11. John, B. et al. Human MicroRNA targets. PLoS Biol 2, e363 (2004).
12. Liang, Y. An expression meta-analysis of predicted microRNA targets identifies a diagnostic signature for lung cancer. BMC Med Genomics 1, 61 (2008).

13. Megraw, M., Sethupathy, P., Corda, B. & Hatzigeorgiou, A. G. miRGen: a database for the study of animal microRNA genomic organization and function. Nucleic Acids Res 35, D149-55 (2007).
14. Vinther, J., Hedegaard, M. M., Gardner, P. P., Andersen, J. S. & Arctander, P. Identification of miRNA targets with stable isotope labeling by amino acids in cell culture. Nucleic Acids Res 34, e107 (2006).
15. Hammell, M. et al. mirWIP: microRNA target prediction based on microRNA-containing ribonucleoprotein-enriched transcripts. Nat Methods 5, 813-819 (2008).
16. Papadopoulos, G. L., Reczko, M., Simossis, V. A., Sethupathy, P. & Hatzigeorgiou, A. G. The database of experimentally supported targets: a functional update of TarBase. Nucleic Acids Res 37, D155-8 (2009).
17. Inomata, M. et al. MicroRNA-17-92 down-regulates expression of distinct targets in different B-cell lymphoma subtypes. Blood 113, 396-402 (2009).
18. The Gene Ontology (GO) project in 2006. Nucleic Acids Res 34, D322-6 (2006).
19. Ogata, H. et al. KEGG: Kyoto Encyclopedia of Genes and Genomes. Nucleic Acids Res 27, 29-34 (1999).
20. Dahlquist, K. D., Salomonis, N., Vranizan, K., Lawlor, S. C. & Conklin, B. R. GenMAPP, a new tool for viewing and analyzing microarray data on biological pathways. Nat Genet 31, 19-20 (2002).
21. Draghici, S. et al. Onto-Tools, the toolkit of the modern biologist: Onto-Express, Onto-Compare, Onto-Design and Onto-Translate. Nucleic Acids Res 31, 3775-81 (2003).
22. Hosack, D. A., Dennis, G., Jr., Sherman, B. T., Lane, H. C. & Lempicki, R. A. Identifying biological themes within lists of genes with EASE. Genome Biol 4, R70 (2003).
23. Volinia, S. et al. A microRNA expression signature of human solid tumors defines cancer gene targets. Proc Natl Acad Sci U S A 103, 2257-61 (2006).
24. Huang da, W. et al. The DAVID Gene Functional Classification Tool: a novel biological module-centric algorithm to functionally analyze large gene lists. Genome Biol 8, R183 (2007).
25. Lee, E. J. et al. Expression profiling identifies microRNA signature in pancreatic cancer. Int J Cancer 120, 1046-54 (2007).
26. Wang, X. & Wang, X. Systematic identification of microRNA functions by combining target prediction and expression profiling. Nucleic Acids Res 34, 1646-52 (2006).
27. Cui, Q., Yu, Z., Purisima, E. O. & Wang, E. Principles of microRNA regulation of a human cellular signaling network. Mol Syst Biol 2, 46 (2006).
28. Gaidatzis, D., van Nimwegen, E., Hausser, J. & Zavolan, M. Inference of miRNA targets using evolutionary conservation and pathway analysis. BMC Bioinformatics 8, 69 (2007).
29. Yoon, S. & De Micheli, G. Prediction of regulatory modules comprising microRNAs and target genes. Bioinformatics 21 Suppl 2, ii93-100 (2005).
30. Gusev, Y., Schmittgen, T. D., Lerner, M., Postier, R. & Brackett, D. Computational analysis of biological functions and pathways collectively targeted by co-expressed microRNAs in cancer. BMC Bioinformatics 8 Suppl 7, S16 (2007).
31. Gusev, Y. Computational methods for analysis of cellular functions and pathways collectively targeted by differentially expressed microRNA. Methods 44, 61-72 (2008).
32. Tagawa, H., Karube, K., Tsuzuki, S., Ohshima, K. & Seto, M. Synergistic action of the microRNA-17 polycistron and Myc in aggressive cancer development. Cancer Sci 98, 1482-90 (2007).
33. Woods, K., Thomson, J. M. & Hammond, S. M. Direct regulation of an oncogenic micro-RNA cluster by E2F transcription factors. J Biol Chem 282, 2130-4 (2007).

34. Taniuchi, K. et al. Overexpressed P-cadherin/CDH3 promotes motility of pancreatic cancer cells by interacting with p120ctn and activating rho-family GTPases. Cancer Res 65, 3092-9 (2005).
35. Kulshreshtha, R. et al. A microRNA signature of hypoxia. Mol Cell Biol 27, 1859-67 (2007).
36. Zarubin, T. & Han, J. Activation and signaling of the p38 MAP kinase pathway. Cell Res 15, 11-8 (2005).
37. Ostrander, J. H., Daniel, A. R., Lofgren, K., Kleer, C. G. & Lange, C. A. Breast tumor kinase (protein tyrosine kinase 6) regulates heregulin-induced activation of ERK5 and p38 MAP kinases in breast cancer cells. Cancer Res 67, 4199-209 (2007).
38. Kim, R., Emi, M., Tanabe, K. & Toge, T. Therapeutic potential of antisense Bcl-2 as a chemosensitizer for cancer therapy. Cancer 101, 2491-502 (2004).
39. Cimmino, A. et al. miR-15 and miR-16 induce apoptosis by targeting BCL2. Proc Natl Acad Sci U S A 102, 13944-9 (2005).
40. Welch, C., Chen, Y. & Stallings, R. L. MicroRNA-34a functions as a potential tumor suppressor by inducing apoptosis in neuroblastoma cells. Oncogene 26, 5017-22 (2007).
41. O'Donnell, K. A., Wentzel, E. A., Zeller, K. I., Dang, C. V. & Mendell, J. T. c-Myc-regulated microRNAs modulate E2F1 expression. Nature 435, 839-43 (2005).
42. Johnson, D. G. & Degregori, J. Putting the Oncogenic and Tumor Suppressive Activities of E2F into Context. Curr Mol Med 6, 731-8 (2006).
43. Jiang, J. et al. Association of MicroRNA expression in hepatocellular carcinomas with hepatitis infection, cirrhosis, and patient survival. Clin Cancer Res 14, 419-27 (2008).
44. Tibshirani, R., Hastie, T., Narasimhan, B. & Chu, G. Diagnosis of multiple cancer types by shrunken centroids of gene expression. Proc Natl Acad Sci U S A 99, 6567-72 (2002).

CHAPTER 10

MATHEMATICAL AND COMPUTATIONAL MODELLLING OF POST-TRANSCRIPTIONAL GENE REGULATION BY MICRORNAS

Raya Khanin[1] and Desmond J. Higham[2]

[1]*Department of Statistics, University of Glasgow,*
Glasgow, G12 8QQ, UK
[2]*Department of Mathematics, University of Strathclyde,*
Glasgow G1 1XH, UK

Mathematical models and computational simulations have proved valuable in many areas of cell biology, including gene regulatory networks. When properly calibrated against experimental data, kinetic models can be used to describe how the concentrations of key species evolve over time. A reliable model allows 'what if' scenarios to be investigated quantitatively *in silico*, and also provides a means to compare competing hypotheses about the underlying biological mechanisms at work. Moreover, models at different scales of resolution can be merged into a bigger picture 'systems' level description. In the case where gene regulation is post-transcriptionally affected by microRNAs, biological understanding and experimental techniques have only recently matured to the extent that we can postulate and test kinetic models. In this chapter, we summarize some recent work that takes the first steps towards realistic modelling, focusing on the contributions of the authors. Using a deterministic ordinary differential equation framework, we derive models from first principles and test them for consistency with recent experimental data, including microarray and mass spectrometry measurements. We first consider typical mis-expression experiments, where the microRNA level is instantaneously boosted or depleted and thereafter remains at a fixed level. We then move on to a more general setting where the microRNA is simply treated as another species in the reaction network, with microRNA-mRNA binding forming the basis for the post-transcriptional repression. We include some speculative comments about the potential for kinetic modelling to contribute to the more widespread sequence and network based approaches in the qualitative investigation of microRNA based gene regulation. We also consider what new combinations of experimental data will be needed in order to make sense of the increased systems-level complexity introduced by microRNAs.

1. Introduction

Most computational efforts to understand post-transcriptional gene regulation by microRNAs (miRNAs) have focused on target prediction tools,[1] and related algorithms, tools and databases are becoming readily available.[2] In this chapter, we deal instead with mathematical and computational modelling of gene regulation by miRNAs. We overview the current state of the art, focusing on our own contribution to the field, and speculate on the role of quantitative modelling in understanding the mechanisms and functions of miRNAs in health and disease. The motivation behind this chapter stems from the need to develop comprehensive models of gene regulation on both transcriptional and post-transcriptional levels with the goal of further integration of such models with target prediction algorithms in the overall complex regulatory network of genes, proteins and RNAs.

Early models for post-transcriptional gene regulation were proposed in,[3–5] and focused on regulation by small RNA (sRNAs). The quantitative two-class model of gene regulation in *E. coli* demonstrated that sRNAs provide a novel mode of gene regulation with a threshold-linear response, a robust noise resistance characteristic, and a built-in capability for hierarchical cross talk.[3] It has also been shown quantitatively that regulation by sRNAs is advantageous when fast responses to external signals are needed and that regulation by sRNA may provide fine-tuning of gene expression.[4] In addition, sRNAs have been suggested to participate in sharpening a gene expression profile that was crudely established by a morphogen.[5]

MiRNAs function very much like small interfering RNAs. However, they are distinguished by their distinct pathway for maturation and by the logic through which they regulate gene expression.[6] Coupled degradation of target mRNA and its regulator[3] is specific to gene regulation by sRNAs. By contrast, miRNAs, which are incorporated into the RISC complex, do not degrade with their targets but return to the cytosol to begin a new round of target mRNA repression. It is plausible, however, that due to increased endonucleolytic activity, miRNA may be degraded after a few cycles of target mRNA binding.[7]

Before going into further detail about kinetic models of gene regulation by miR-NAs, we outline some significant experimental findings that guide the modelling process. It is well established that miRNAs regulate gene expression post-transcriptionally, influencing stability, compartmentalization and translation of mRNAs (see Figure 1). The underlying molecular mechanisms are still debated[8] but the overall effect of miRNA appears to be repressive. It has been computationally predicted[9] and demonstrated experimentally[10–12] that one type of miRNA may regulate large number, sometimes in the hundreds, of different types of target mRNAs and proteins.

The regulating effect of miRNAs is typically studied by conducting miRNA mis-expression experiments and measuring gene expression with microarrays,[10] protein levels by the powerful mass spectromic method SILAC[11] and changes in protein synthesis by pulsed SILAC (pSILAC).[12] There are two general types of miRNA mis-expression experiments,[11–13]

(1) overexpression or transfection of a miRNA to a cell-line or tissue where it is initially not present, or present at small levels, and
(2) knock-down of an abundant miRNA by antagomir or LNA.

Recent proteomic and microarray studies have isolated the major sequence determinant that mediates miRNA regulation of both mRNAs and protein: the 6-mer "seed" (Watson-Crick consecutive base pairing between mRNAs and the miRNA at position 2-7 counted from its 5' end) located in the 3' untranslated regions (3'UTRs) of mRNAs.[12] The accuracy of target prediction algorithms can be improved by allowing for evolutionary conservation of the seed site.[12] However, the false-positive rate of target predictions, even with conserved seed incorporation, is still estimated at 40% (using the entire pSILAC dataset). The extent of the miRNA-mediated regulation is relatively mild and depends on the number of seeds and the distances between them. It also is affected by by many other factors, including the sequence elements around the seed, sites for RNA-binding proteins and secondary structure. This is a subject of intensive research. Availability of mRNA expression levels from miRNA mis-expressions and various tissues as well as protein levels as measured by SILAC[11] and protein production as measured by pSILAC[12] gives an opportunity to improve existing target prediction algorithms.[14]

MiRNAs are thought to act by binding to their target mRNAs rather than through a catalytic mechanism requiring only a transient association.[15] The bulk of miRNAs are linked with target mRNAs undergoing translation.[16] It is reasonable to assume that the miRNA:mRNA complexes are being translated, but at a slower rate than free mRNAs. Certain mRNAs accumulate in P-bodies (PBs) in a miRNA-dependent manner,[17] raising the possibility that PBs might be involved in miRNA-mediated repression. However, miRNA-mediated repression is unaffected in cells devoid of microscopically visible PBs, suggesting that PB formation itself is not required for repression. This led to the conclusion that PBs are a consequence, not the cause, of the miRNA-mediated silencing.[18]

Just as there are unanswered questions relating to the sequence characteristics of each miRNA:mRNA pair, relatively little is known about how the extent of regulation depends on the expression levels of the miRNA and the target of interest, as well as other targets that are affected by that miRNA. This is an area where modelling and computation hold great promise for adding value to experimental data sets. MiRNA mis-expression data can be used to infer kinetic parameters for various processes involved and to elucidate the mechanisms of miRNA-mediated target regulation. Other problems that can be addressed include estimating the time-response of the system to miRNA transfection and knock-down (de-repression). As more data sets, including miRNA studies by proteomics, become available, there is further potential for dynamical models to add a quantitative level of understanding that can be used to test hypotheses and tackle 'what if' scenarios.

2. A Simple Model of miRNA-Mediated Gene Regulation

We now discuss a simple yet plausible model of target regulation by miRNA. The model summarizes the overall effect of miRNA on a target, without representing

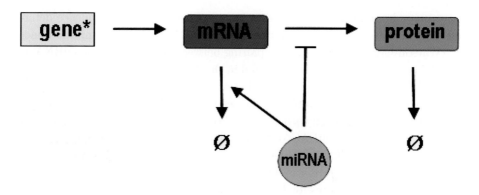

Fig. 1. MicroRNA exerts its downregulating effect on mRNA or/and protein.

specific molecular mechanisms of the downregulation. This model assumes that the miRNA level remains fixed—see section 5 for a discussion of the more general case where kinetics for miRNA is introduced.

Gene expression, or mRNA levels, m, and protein levels, p, change through production and degradation.[19] We may describe these rates of change with an ordinary differential equation (ODE)

$$\frac{dm(t)}{dt} = q(t) - \delta\, m(t), \tag{1}$$

$$\frac{dp(t)}{dt} = \lambda\, m(t) - \delta_P\, p(t). \tag{2}$$

In models of transcriptional regulation it is commonly assumed that while the production rate governing mRNAs, $q(t)$, depends on the availability of a single, or multiple, transcription factors (TFs), the mRNA degradation and protein translation are first-order processes that occur with constant rates, δ and λ, respectively. We also assume a constant protein degradation rate, δ_P.

When a transcript is a target of a specific miRNA, its degradation rate, δ, and its rate of translation, λ, both depend on the levels of this miRNA: $\delta = \delta(\text{miRNA})$ and $\lambda = \lambda(\text{miRNA})$. It is known that the presence of miRNA enhances the degradation rate[10,13] and inhibits translation[11,12,20,21] (Figure 1).

The mRNA target degradation rate may be written as

$$\delta(\text{miRNA}) = \delta_0 \left(1 + d(\text{miRNA})\right), \tag{3}$$

where $d(\text{miRNA})$ represents the miRNA-mediated fold-change in the target mRNA degradation rate relative to the basal degradation rate, δ_0. Plausible forms include linear (mass-law)[22]

$$\delta(\text{miRNA}) = \delta_0 \left(1 + d\,\text{miRNA}\right) \tag{4}$$

and Michaelis-Menten type dependency

$$\delta(\text{miRNA}) = \delta_0 \left(1 + d \frac{\text{miRNA}}{\gamma + \text{miRNA}}\right). \quad (5)$$

To account for multiple seeds for the same miRNA,[1] the above models are easily extended to include a cooperativity, or Hill, coefficient, with

$$\delta(\text{miRNA}) = \delta_0 \left(1 + d \cdot \text{miRNA}^h\right), \quad (6)$$

$$\delta(\text{miRNA}) = \delta_0 \left(1 + d \cdot \text{miRNA}^h / (\gamma + \text{miRNA}^h)\right), \quad (7)$$

replacing (4) and (5), respectively. Here $h \geq 1$ is the number of multiple seeds for the same miRNA with optimal spacings of ≤ 40 nt.[12,23]

The miRNA-mediated translational repression in (2) can be described by

$$\lambda(\text{miRNA}) = \frac{\lambda_0}{\gamma + a\,\text{miRNA}^h}, \quad h \geq 1, \quad (8)$$

where $a = 1$ when mRNA is a translational target of this specific miRNA and $a = 0$ otherwise.

One miRNA may have many target genes.[9–12] The expression level of the ith target mRNA transcript i may then be described by the ODE:[24]

$$\frac{dm_i(t)}{dt} = q_i - \delta_{0i}\left(1 + d_i(\text{miRNA})\right)m_i(t),$$

$$\frac{dp_i(t)}{dt} = \lambda_i(\text{miRNA})m_i(t) - \delta_{Pi}\,p_i(t).$$

Kinetic parameters $q_i, \delta_{Pi}, \delta_{0i}$ are gene-specific. In addition, parameters d_i, γ_i, and h_i of the miRNA-mediated downregulation that appear in the functional relations $d_i(\text{miRNA}) \geq d_i(\text{miRNA} = 0)$ and $\lambda_i(\text{miRNA}) \leq \lambda_i(\text{miRNA} = 0)$ are also target specific and depend on the sequence and structure of each mRNA:miRNA base pairing and other recognition elements that have not yet been identified.

The logic of transcriptional regulation by TFs and post-transcriptional regulation by miRNAs appears to be the same.[25] Two, or more, miRNAs can regulate a target in different ways. Suppose that two miRNAs, miRNA$_1$ and miRNA$_2$, regulate a target i. Then, a SUM-type of regulation may be represented by

$$\delta_i = \delta_{0,i}\left(1 + d_{i,1}(\text{miRNA}_1; h_{i,1}) + d_{i,2}(\text{miRNA}_2; h_{i,2})\right), \quad (9)$$

$$\lambda_i = \frac{\lambda_{0i}}{1 + \text{miRNA}_1^{h_{i,1}}/\gamma_{i,1} + \text{miRNA}_2^{h_{i,2}}/\gamma_{i,2}}. \quad (10)$$

This is a non-cooperative regulation, wherein the effect of two miRNAs is simply additive. A cooperative SUM-regulation takes place if the presence of two miRNAs results in larger regulation than the sum of each miRNA-mediated regulation. An example of cooperative SUM logic is demonstrated in,[23] wherein the downregulation of mRNAs in overexpression experiments is larger if the distance between the seeds for transfected miRNA and one of endogenous miRNAs is within an optimal range. Alternatively, an AND gate logic is relevant if both miRNAs must be present:

$$\delta_i = \delta_{0,i}\Big(1 + d_{i,1}(\mathtt{miRNA}_1; h_{i,1})\, d_{i,2}(\mathtt{miRNA}_2; h_{i,2})\Big), \tag{11}$$

$$\lambda_i = \frac{\lambda_{0,i}}{1 + \mathtt{miRNA}_1^{h_{i,1}}\, \mathtt{miRNA}_2^{h_{i,2}}/\gamma_i}.$$

Other types of regulation, such as non-exclusive or exclusive OR gates, are also possible and can be described by similar types of equations.

2.1. Fold-changes of mRNA and proteins at different miRNA levels

The silencing effect of miRNA is estimated experimentally by comparing levels of mRNAs and proteins at two different miRNA levels, usually when miRNA is present ($\mathtt{miRNA} > 0$) or it is absent($\mathtt{miRNA} = 0$). Microarray and proteomic measurements are taken in miRNA mis-expression experiments or across different tissues, cell-lines or conditions. The steady state levels of target mRNAs and proteins ($i = 1\ldots N$) at a given level of the $\mathtt{miRNA}$ can readily be written as

$$m_i = \frac{q_i}{\delta_{0,i}(1 + d_i(\mathtt{miRNA}))} \quad \text{and} \quad p_i = \frac{\lambda_i(\mathtt{miRNA})m_i}{\delta_i^p}.$$

Microarrays, SILAC[11] and pSILAC[12] measure fold-changes in mRNAs, protein levels or the number of newly produced proteins under different conditions. The fold-changes for mRNA and protein i are given by

$$FC_i^{\mathrm{mRNA}} := \frac{m_i}{m_i(\mathtt{miRNA}=0)} = \frac{1}{1 + d_i(\mathtt{miRNA})} \leq 1 \tag{12}$$

$$FC_i^{\mathrm{prot}} := \frac{p_i}{p_i(\mathtt{miRNA}=0)} = \frac{\lambda_i(\mathtt{miRNA})}{\lambda_i}\frac{m_i}{m_i(\mathtt{miRNA}=0)}$$

$$\leq \frac{m_i}{m_i(\mathtt{miRNA}=0)} < 1. \tag{13}$$

The inequality (13) predicts that downregulation of proteins is larger than that of mRNAs, as has indeed been observed for the majority of targets.[12] It follows immediately from (12) and (13) that

$$\log_2 FC_i^{\mathrm{prot}} = \log_2 FC_i^{\mathrm{mRNA}} + \Lambda_i, \tag{14}$$

where

$$\Lambda_i = \log_2[\lambda_i(\mathtt{miRNA})] - \log_2[\lambda_i] < 0.$$

Here Λ_i represents direct miRNA-mediated repression of translation that was computed from experimental data by subtracting the log2 mRNA from the log2 pSILAC fold-changes.[12] Using model (8) for miRNA-regulated rate of translation, Λ_i can be rewritten as

$$\Lambda_i = -\log_2[1 + \frac{\mathtt{miRNA}^{h_i}}{\gamma_i}],$$

where h_i is the number of seeds acting cooperatively. If miRNA action on the target i results in relatively large translational repression, so that $\mathtt{miRNA}_i^h/\gamma_i \gg 1$, then

$$\Lambda_i \approx -\log_2[\frac{\mathtt{miRNA}^{h_i}}{\gamma_i}] = -h_i \log_2[\mathtt{miRNA}] + \log_2[\gamma_i]. \qquad (15)$$

It follows from the above formula that, for large negative fold-changes, translational repression measured as log fold-change, is linearly correlated with the number of seeds, as has indeed been observed by comparing pSILAC and microarray data.[12] The average number of seeds, plotted as a function of the differences between protein and mRNA fold-changes, exhibits linear decay towards the regime of equal fold-changes (Figure 5c,d;[12]), indicating that in addition to mediating mRNA downregulation, the seed also mediates direct repression of translation rates for hundreds of genes.

It was reported that the slope of the average number of seeds is steeper for pSILAC fold changes than for mRNA fold-changes, suggesting that the multiplicity of a miRNA-binding site in the same 3'UTR exerts a stronger direct effect on protein production than on mRNA levels.[12] This observation is consistent with (15), since Λ_i is directly proportional to h_i.

We have shown that the above formulas reproduce experimental observations of.[12] They also suggest that the effect of both mRNA and translational repression has a weak (logarithmic) dependence on the level of miRNA itself. To verify this prediction would require further experimental data from miRNA mis-expression experiments with different levels of miRNA. Our modelling also predicts that fold-changes in mRNA and protein levels do not depend on the initial target level. So targets that are expressed at different levels are downregulated to the same extent provided the kinetic parameters of the miRNA regulation are the same. Indeed, the average protein fold-changes (pSILAC data from the Ref.[12]) for targets with seeds in their 3'UTRs as functions of their mRNA levels at control do not exhibit any pattern in protein fold-changes for a large range of control intensities [24, Figure 2].

3. Non-Steady-State Behaviour

It is also informative to consider non-steady state behaviour of the model (1)–(2). Fitting the model to temporal microarray and pSILAC (or SILAC) data upon miRNA transfection/knock-down will yield kinetics of the miRNA-mediated effect on a target gene's degradation and translation rates. Indeed,[22] demonstrated that it is possible to infer kinetic parameters of miRNA-mediated mRNA degradation using temporal gene expression data from overexpression experiments. Here we explore scenarios of miRNA mis-expression experiments wherein miRNA levels quickly change to a new constant level due to transfection or knock-down.

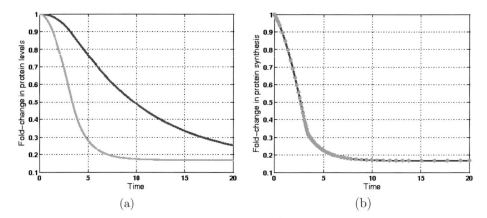

Fig. 2. Protein profiles post-transfection with miRNA: (a)-(b) miRNA experiment: "red" protein has degradation rate that is ten-fold larger than that of "green" protein.

3.1. *MicroRNA levels are constant*

If the miRNA level in a cell-line or tissue does not change much with time post-transfection (or post-knock-down), the value of miRNA in the system (1)–(2) can be fixed at some arbitrary level and a closed-form solution for each target mRNA for the transfection experiment can readily be found:

$$m_i(t) = m_i^0 \, e^{-\delta_i(\texttt{miRNA})t} + m_i(\texttt{miRNA}) \left(1 - e^{-\delta_i(\texttt{miRNA})t}\right),$$

$$p_i(t) = p_i^0 \, e^{-\delta_{pi}t} + p_i(\texttt{miRNA}) \left(1 - e^{-\delta_{pi}t}\right)$$
$$+ \frac{\lambda(\texttt{miRNA}) \, q_i}{\delta_{0i} \, \delta_i(\texttt{miRNA})} \frac{\delta_i(\texttt{miRNA}) - \delta_{0i}}{\delta_{pi} - \delta_i(\texttt{miRNA})} \left(e^{-\delta_i(\texttt{miRNA})t} - e^{-\delta_{pi}t}\right),$$

where the initial levels of target mRNA and protein i are given by

$$m_i^0 = m_i(\texttt{miRNA} = 0) = \frac{q_i}{\delta_{0i}}, \quad p_i^0 = p_i(\texttt{miRNA} = 0) = \frac{\lambda_i}{\delta_{pi}} m_i^0$$

and the steady states at the new miRNA level are

$$m_i(\texttt{miRNA}) = \frac{q_i}{\delta_i(\texttt{miRNA})}, \quad p_i(\texttt{miRNA}) = \frac{\lambda_i(\texttt{miRNA})}{\delta_{pi}} m_i(\texttt{miRNA}).$$

The above equations can be used for estimating time-responses of different targets to miRNA transfection. It is clear that the time-scale for introducing miRNA-mediated repression is determined by either δ_{pi} or $\delta_i(\texttt{miRNA})$, whichever is slower. Therefore, the levels of high-turnover proteins (large δ_{pi}) will change rapidly whereas stable proteins (low δ_{pi}) will be affected later, as seen in Figure 2. If the measurements of protein levels by standard techniques such as SILAC are done at times less than 10–15h post-transfection, then erroneous conclusions will be reached that the "red" protein is downregulated to a lesser degree than the "green" protein. Thus, to assess endogenous regulation of mRNA translation by miRNAs, a technique such as pulsed SILAC is needed to measure directly genome-wide changes in protein synthesis shortly after changes in miRNA expression.[12]

Similarly, the time-course of de-repression (removing miRNA-mediated repression) can be found by solving equations (1)–(2) with knocking-down miRNA at $t = 0$ (miRNA = 0) and initial conditions mRNA(0) = m_i(miRNA) and prot(0) = p_i(miRNA). The time-responses for different types of mRNAs and proteins can be estimated and compared with removing TF-mediated repression, similar to the case of small RNAs as has been done by.[4] The simple model (1)–(2) adequately describes experimental observations from miRNA transfections and knock-downs as shown above. This model, however, does not include miRNA-mediated relocation of mRNAs into P-bodies that can quickly be released into the cytoplasm.[17] This suggests that to account for quick miRNA-mediated repression and de-repression,[25] as occurs when the miRNA profile drastically changes in a short time-window in development processes or in response to stress, the step of relocation/release of miRNA:mRNA complexes from P-bodies and their subsequent translation is required.

3.2. MicroRNA levels are changing

It is quite plausible that miRNA levels are influenced by TFs,[26] or by other miRNAs.[27] In addition, miRNAs can decay before they are incorporated into the stable RISC complexes or degrade in the process of a few cycles of mRNA binding. The latter can be accounted by introducing a global parameter that represents the probability for a miRNA to be co-degraded with the mRNA in the processed state. Additionally, miRNA molecules, incorporated into the RISC complexes, are sequestered into the target mRNA:miRNA complexes, subsequently translocated to the P-bodies. This can be modelled as a multi-step process, as discussed in section 5. Some of the mRNA:miRNA complexes might be stored in the P-bodies to be released in response to stress.[17] In miRNA overexpression experiments, the miRNA levels can decrease due to cell growth/division, and natural miRNA decay. The effects that cause miRNA levels to decrease can be lumped together and considered as a linear decay process of miRNA with the rate δ_m:[22]

$$\frac{d\text{miRNA}(t)}{dt} = s(t) - \delta_m \text{miRNA}(t). \tag{16}$$

Here $s(t)$ is the rate of miRNA transcription that might depend on various TFs. In cell-lines and tissues where a specific miRNA is present, its level can be approximated by s/δ_m. In miRNA transfection experiments, the miRNA level is maximum at the initial (transfection) time: miRNA(0) = 1. Assuming no production of the miRNA in the cell-line or tissue where it is normally not present, $s(t) = 0$, the miRNA temporal profile is determined by just one parameter, δ_m: miRNA$(t) = e^{-\delta_m t}$. Interestingly, there is an experimental observation that the majority of downregulated target mRNA expression profiles measured for up to 120h in miR124a transfection show upregulation in the last time-point(s),[28] perhaps indicating that the subsequent upregulation of the target mRNAs is due to a decrease in the level of the transfected regulator miRNA.[22]

3.3. Reconstructing kinetics of mRNA-regulated Single Input Motif from high-throughput data

Kinetic parameters of miRNA-mediated target downregulation are difficult to measure directly, but they can be inferred via time-course high-throughput datasets from miRNA mis-expression experiments using a minimal model (1)–(2). The first attempt in this direction has been done in the Ref.,[22] where the authors apply the model to temporal gene expression data from a miRNA transfection experiment.[28] Similarly, kinetics of miRNA-mediated translational repression can be estimated from protein measurements.

Let us consider the simplest miRNA-regulated gene circuit that involves one miRNA that post-transcriptionally acts on its numerous targets. This structure is similar to the so-called Single Input Motif (SIM) that is common in transcription networks.[29] A miRNA-SIM has larger number of targets on average than a TF-SIM. Both types of SIMs are condition-specific because many targets are presumably subject to control by other miRNAs or TFs whose levels do not change in the course of the experiment. Their effects are implicitly incorporated in the basal rate-constants of transcription and degradation of target mRNAs and proteins.

Kinetic parameters of the SIM targets $\{q_i,\ \delta_{i0},\ \lambda_i,\ \delta_{pi},\ \delta_i(\texttt{miRNA}),\ \lambda(\texttt{miRNA})\}$, $i = 1\ldots N$, as well as the unknown profile of the miRNA master regulator that is common to all targets, can be reconstructed from high-throughput data.[30] One way of estimating parameters of miRNA-mediated mRNA degradation from microarray data involving miRNA overexpression is by the maximum likelihood procedure.[22] This method applied to time-course post-transfection with miRNA124a yielded a miRNA124a half-life of $29h$ (95% confidence bounds $(26h, 50h)$). This estimate incorporates the effects of free miRNA decay, its sequestering into P-bodies as well as cell growth (and thereby miRNA dilution). In addition, it has been observed in the study[22] that the miRNA downregulating effect on the target mRNA degradation rates can adequately be described by either a non-linear (5) or a linear model (4). In order to distinguish between the two models, miRNA mis-expression (overexpression and silencing) experiments are required where the miRNA levels can be measured. A model that takes into account multiple sites for the same miRNA on the 3'UTR of the target mRNAs (6), gives a better fit to some mRNA profiles, so the number of active seeds, h, can also be estimated from the data. Experiments wherein the mRNA expression is measured at different levels of miRNA would be particularly helpful in determining the miRNA dosage-dependent effect on the target downregulation.[22] estimated effective miRNA-mediated fold-change increase in each target mRNA degradation rate, and the reconstructed basal decay rates of target mRNAs in this study have a very good correspondence with experimental measurements from an independent study,[31] thereby giving a strong support for this modelling approach. These methods, with extended modelling assumptions, and other optimization and inference techniques, such as Bayesian inference,[32] can be applied to new experimental datasets and will yield kinetic information on miRNA-regulation, as well as miRNA time-course and biogenesis.

4. Models of miRNA-Mediated Network Motifs

As different types of high-throughput data accumulate, it is natural to integrate them together with experimentally verified regulatory relationships as well as those obtained from bioinformatics tools and existing functional genomics data.[1] An initial effort in this direction has been published for nematodes.[33] There is now a growing number of papers that explore the "wiring" of miRNA regulatory relationships together with known transcriptional regulatory interactions,[26,34,35] signal transduction networks[36] and protein-protein interactions networks.[37,38] These networks are constructed by using computational predictions for miRNA targets and transcription factor binding sites. It appears that a very large number (up to 43%) of human genes are under combined transcriptional and post-transcriptional regulation.[26] The true number of human genes that are subject to a dual TF and miRNA regulation is probably even higher considering the fact that the collection of mammalian miRNAs is not yet complete.

Studies of network structures that involve miRNAs and TFs started by considering just a few experimentally confirmed cases.[39] Some network motifs have been experimentally found in *C.elegans*, notably a double-negative feedback loop.[40] More recently, composite feedback loops in which a TF that controls a miRNA is itself regulated by that same miRNA have been shown to be over-represented.[41] TFs that control cell proliferation and apoptosis (Myc and E2F), and crucial pathways, such as the p53 master network, have been found to be under tight control of several miRNAs.[42,43] The goal now is to identify the over-represented motifs in large networks and to understand how functionality is related to structure.[26,35,44]

Initial kinetic models of gene regulatory circuits with a feedback between miRNAs and TFs have recently appeared.[35,42,45] In these studies, a TF is one of the targets of the miRNA under consideration, whose transcription rate, $s(t)$ (see equation 16) is regulated, positively or negatively, by the same TF (here denoted by P with cooperativity n):

$$s(t) = \beta \frac{P^n}{(\gamma + P)^n} \text{ for } activator; \quad s(t) = \frac{\beta}{(\gamma + P)^n} \text{ for } repressor. \qquad (17)$$

Mathematical models of the basic miRNA-TFs structures demonstrate their complex and intricate behaviour. The TF-miRNA feedback circuit can operate as the simplest biological switch,[45] potentially changing levels of a large number of other targets. Feedback loops often include autoregulation of TF (e.g. a cancer network that comprises miR-17-92 cluster, E2F and Myc[42]). Such autoregulated feedback loops exhibit changes in the steady-state levels of TF and miRNA that go in the same direction.[42] This agrees with experimental observations on Myc in miR-17-92 cluster in various tumors, but is somewhat counterintuitive, as one might expect that a mRNA/protein that is a target of a miRNA and the miRNA are expressed reciprocally in different tissues—this has indeed been shown to be the case for many targets.[46]

The case of two TFs regulating each other with one miRNA regulating both of them has been shown to be the most significant overrepresented network motif in a human regulatory network.[35] Mathematical modelling demonstrated that miRNA stabilizes mutual regulation of two TFs to resist perturbations. On the other hand, such a motif has the ability to convert a transient stimulus into a stable and irreversible response. Mathematical modelling of the fundamental structures has already demonstrated that the basic repressive function of miRNAs when combined with other regulatory factors can build up more complex and higher-order capabilities such as fine-tuning, canalization[39] and multi-dimensional switching.

It must be noted, however, that current models of miRNA-TF feedback circuits do not take into account the potential impact of numerous other miRNA targets on the circuit behaviour.[35,42] More realistic models of such circuits should be extended to include the potential effect of multiple miRNA targets.[24,45] Full models that study functionalities of miRNA-TF network motifs should also take into account all possible modes of miRNA-mediated target regulation, i.e. mRNA destabilization and/or repression of translation, and multiple seeds for a miRNA (cooperativity in equation (6)). According to a recent study,[35] there exist two classes of miRNAs: miRNAs in one class are regulated by TFs, while the other class of miRNAs regulate TFs. Modelling the behaviour of both miRNA classes requires detailed consideration of how inputs from different types of regulators, miRNAs or TFs, exert their regulatory effect (equations (9) and (11)). Another feature of the TF and miRNA regulatory circuits that may be significant is the time delay between production and regulatory action.[42,47]

Modelling network structures is crucial in unravelling miRNA functionality and its relationship with other regulatory factors. Related questions include whether miRNA-driven switches or double-negative feedback loops[40] have any advantage over similar structures that are controlled on transcriptional level. TF-driven switches and double-negative feedback loops have been theoretically and computationally studied in great detail by several authors. Similar studies of basic miRNA-mediated switches will undoubtedly be performed in the near future, and will reveal the commonalities and differences between these structures. It seems likely that interwoven regulatory combinations of miRNAs and TFs yield robust multi-dimensional switches that frequently occur in differentiation (RK, unpublished observations).

In the next section we discuss how the simple model (1)–(2) can be extended to include the binding of miRNA to its target mRNAs.[7,24] This brings about miRNA-mediated *cross-talk* if miRNA degrades after a few cycles of target mRNA binding.

5. A Multi-Step Model: Including miRNA Binding to Target mRNA

The two-step model of the Ref.[7] is based on plausible biological assumptions: the binding of miRNA to the mRNA promotes a secondary process (e.g. ribosome run-off or deadenylation) that ultimately leads to mRNA accumulation in its processed state, perhaps in P-bodies. The authors show that the target mRNA and protein levels may be tuned by target-specific parameters while global effectors may alter this behavior for some, but not all, miRNA targets in the cell. However, this model erroneously predicts that the fold-changes in protein levels can not be higher than those of corresponding mRNAs (equation 2 in the Ref.[7]), contrary to experimental findings.[12] This is due to their assumption that proteins are produced at equal rates from both free and miRNA-bound mRNAs.

A model that takes into account the miRNA binding to mRNAs has recently been developed.[24] The model yields results that are consistent with miRNA misexpression experiments. Here we will outline the model assumptions, display the ODE system, and briefly present the results.

To set up a model, consider a system where a single type of miRNA targets several different types of mRNA molecules, m_i. We assume that each type of mRNA is produced with its own transcription rate q_i and decays with its own rate δ_i.

The miRNA itself is being produced in the cell with a rate p_m and decays with a rate δ_m. The model allows mRNA and miRNA to form a complex, $\texttt{miRNA} \cdot m_i$,

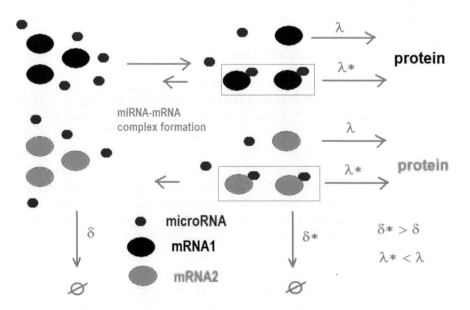

Fig. 3. Cartoon of microRNA bindings to mRNA targets.

with a forward rate β_i and a reverse rate β_i^-. The complex $\texttt{miRNA} \cdot m_i$ decays at a rate $\delta_i^\star$. Proteins, p_i, are being translated at a rate λ_i from free mRNAs, m_i, and with a rate $\lambda_i^\star$ from the complexes, $\texttt{miRNA} \cdot m_i$, and degrade at a rate δ_i^p. The key downregulating property of the miRNA is introduced by two constraints:

- $\delta_i^\star \geq \delta_i$, so that the complex degrades faster than free mRNA, and/or
- $\lambda_i^\star \geq \lambda_i$, so that the complex produces protein more slowly than free mRNA.

The ratios $\delta_i^\star/\delta_i$ and $\lambda_i^\star/\lambda_i$ will depend on specific target mRNA and miRNA base-pairing in and around the seed region. We further suppose that when the complex degrades, a fixed proportion, $0 \leq \kappa \leq 1$, of the miRNA returns to the pool.

In the case where there are two targets, the ODE model may be written

$$\frac{dm_1}{dt} = q_1 - \delta_1 m_1 - \beta_1 m_1 \texttt{miRNA} + \beta_1^- \texttt{miRNA} \cdot m_1$$

$$\frac{dp_1}{dt} = \lambda_1 m_1 - \delta_1^p p_1 + \lambda_1^\star \texttt{miRNA} \cdot m_1$$

$$\frac{dm_2}{dt} = q_2 - \delta_2 m_2 - \beta_2 m_2 \texttt{miRNA} + \beta_2^- \texttt{miRNA} \cdot m_2$$

$$\frac{dp_2}{dt} = \lambda_2 m_2 - \delta_2^p p_2 + \lambda_2^\star \texttt{miRNA} \cdot m_2$$

$$\frac{d\texttt{miRNA}}{dt} = p_m - \delta_m \texttt{miRNA} - \beta_1 m_1 \texttt{miRNA} + \beta_1^- \texttt{miRNA} \cdot m_1$$
$$- \beta_2 m_2 \texttt{miRNA} + \beta_2^- \texttt{miRNA} \cdot m_2 + \delta_1^\star q \texttt{miRNA} \cdot m_1 + \delta_2^\star q \texttt{miRNA} \cdot m_2$$

$$\frac{d\texttt{miRNA} \cdot m_1}{dt} = \beta_1 m_1 \texttt{miRNA} - \beta_1^- \texttt{miRNA} \cdot m_1 - \delta_1^\star \texttt{miRNA} \cdot m_1$$

$$\frac{d\texttt{miRNA} \cdot m_1}{dt} = \beta_1 m_1 \texttt{miRNA} - \beta_1^- \texttt{miRNA} \cdot m_1 - \delta_1^\star \texttt{miRNA} \cdot m_1.$$

This generalizes readily to any number of targets.[24]

Although this nonlinear ODE system cannot be solved analytically, it is possible to analyse the steady state behaviour. It can be shown that the $\kappa = 1$ regime completely uncouples the targets, while $\kappa < 1$ introduces coupling into the system, wherein the level of achieved miRNA-mediated repression is dependent on the amount of miRNA itself, the expression level of the target under consideration and levels of other targets.[15] Indeed, miRNA degradation after several rounds of binding to target mRNA results in miRNA-mediated target cross-talk, when changing the level of one target has an effect on the level of other target(s). An increase in the level of a target can be caused by an external signal, or by a feedback in the circuit itself, wherein one of the targets is a transcription factor that regulates the other one. Additionally, as discussed above, the feedback between target TFs in the circuit and miRNA can change the levels of the miRNA itself.

To illustrate these effects, Figure 4 presents numerical simulations of the full ODE model. We show output over a time interval $0 \leq t \leq 10$, where at time

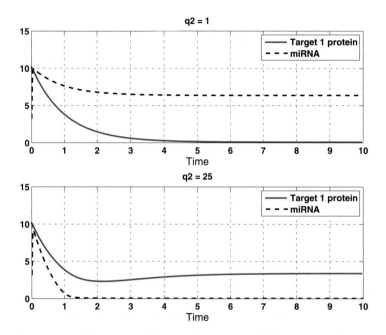

Fig. 4. Levels of target one protein and miRNA, from the full ODE model. In the lower picture, where $q_2 = 25$, target two mRNA production is much greater and target one protein production has indirectly benefited. Parameter values $q_1 = 5$, $\delta_1 = 1$, $\delta_2 = 1$, $b_1 = 50$, $b_2 = 50$, $b_1^- = 0.1$, $b_2^- = 0.1$, $\lambda_1 = 2$, $\lambda_2 = 1$, $\delta_1^+ = 1$, $\delta_2^+ = 1$, $\lambda_1^\star = 2$, $\lambda_2^\star = 2$, $p_m = 10$, $\delta_m = 1.1$, $\delta_1^\star = 100$, $\delta_2^\star = 100$ and $\kappa = 0.5$. Note that this effect is a consequence of the *relative* sizes of the parameters. The absolute values were chosen arbitrarily and are not intended to match any particular biological system.

zero, each species was set to a level of 10. The figure 4 shows the level of target one protein $p_1(t)$, and the level of miRNA(t). We set $q_2 = 1$ in the upper picture and $q_2 = 25$ in the lower picture. We see from the figure that increasing q_2, that is, raising the transcription rate for the second target, causes the level of target one protein to rise. This is explained by the decreased availability of miRNA. The repressive effect of miRNA on target one has been reduced because more miRNA is binding to the over-abundant target two.

The figure illustrates that levels of mRNAs that are targets of the same miRNA are interdependent, via the level of miRNA that affects them. If the levels of one (or several) mRNA(s) increase(s) due to transcriptional control, the miRNA may become limiting and its downregulating effect on the rest of the targets will be substantially decreased. If it is crucial to keep other targets at a low level, then additional controls at either transcriptional or post-transcriptional level (by combinatorial regulation with another miRNA) are needed. Indeed, a recent study strongly suggested that coordinated transcriptional and post-transcriptional (p53 and miRNAs mediated) networks are an integral part of tumorigenesis.[43]

Extrapolating from our simple illustration of two targets to the case where several types of miRNA interact with multiple targets, it becomes clear that the problem of inferring regulatory relationships directly from mRNA expression data alone will generally be infeasible due to the high level of cross-talk.

For a large number of targets, the multi-step model involves many different parameters. It is yet to be determined whether rate-constants for miRNA-mRNA complex formation/dissociation are target specific or whether it is plausible to assume that the kinetics of the translocation to/from P-bodies is governed by global parameters that are determined by cell condition.[7]

We emphasize that the above multi-step model was not designed to address the molecular mechanism of miRNA action. A very promising step in this direction has recently been undertaken by,[48] who examined by means of computational analysis the effect of miRNA on different steps in translation. Kinetic analysis of their model with rate constants carefully estimated from experimental data demonstrates that a miRNA will have a modest effect on the overall rate of protein production from a specific target mRNA if the step it affects is not rate-determining. Their model is consistent with the suggestion that miRNAs may primarily repress translational initiation at a late step. However, the authors demonstrate that the experimental observations used to argue for this suggestion are open for alternative interpretations.[48]

6. Future Prospects

In this review, we described recent progress in computational and mathematical modelling of post-transcriptional gene regulation by miRNAs. We can divide current approaches in miRNA Systems Biology into three groups. The first approach deals with computational prediction of miRNA targets. A growing number of algorithms predict miRNA sites on the targets taking into account miRNA-mRNA base-pairing and its secondary structures, seed conservation among species and other sequence characteristics, including AU-richness and proximity of RNA motifs. Despite various underlying and additional modelling assumptions, all target prediction tools are clearly **sequence**-based.

Another approach that is gaining momentum is integration of available predictions from the **sequence**-based methods together with other regulatory relationships, such as transcription factors and their targets, into a **network**-based approach. Unlike purely transcriptional regulatory networks studied by many groups, miRNA–TF networks have an additional dimension, as each target (node) can be regulated on two different levels: transcriptional and post-transcriptional. In addition, regulators in this network, TFs and miRNAs, mutually regulate each other. Moverover, post-transcriptional regulation involves miRNA-mediated destabilization of target mRNA and/or translational repression, and the links connecting the two would ideally indicate the mode of regulation. Study of two-dimensional, directed networks clearly requires new tools for description, representation, visualization and efficient search algorithms for over-represented motifs.

In this chapter, we focused on a third class of models, based on `expression` levels. Although we restricted ourselves to kinetic modelling of miRNA-mediated gene regulation using ordinary differential equations, we note that equations can easily be translated into stochastic form in order to study extrinsic and intrinsic noise and the cases of low numbers of molecules.[49]

With very few exceptions, `sequence`, `network` and `expression` approaches used in computational studies of miRNA-mediated regulation currently go pretty much in parallel without much cross-talk between them. Clearly, both `network` and `expression` approaches require predictions from `sequence` algorithms. In addition, ODE models of over-represented network motifs are being built and studied. But can the results from the `network` and ODE based approaches improve target prediction tools? In our view, they can. Future experimental data should be interpreted by taking into account not only sequences of the participating mRNAs and proteins, as has been successfully done in the past.[11,12] Target prediction and verification can benefit from including information on expression levels of miRNAs and mRNAs as well as details of regulatory miRNA–TF structures. On one hand,[24] have shown theoretically and verified on pSILAC datasets that targets that are expressed at different levels are downregulated to the same extent (i.e. their fold-changes are equal) provided the kinetic parameters of the miRNA regulation are the same. On the other hand, the number of false-positives in pSILAC datasets (mRNAs with seeds that are not downregulated) in lowly expressed mRNAs at control is likely to be considerably larger than for those mRNAs that are expressed at higher levels (RK, unpublished observations on pSILAC data).

Dynamic modelling of network structures and pathways from a consortium of datasets that will become available in the next 5–10 years might therefore point at some additional features that ought to be included in target prediction algorithms. An important contribution would be to identify specific conditions when certain miRNA sites are active, as they require cooperative action from other nearby miRNA sites or RNA-binding proteins, thereby reducing the fraction of false-positives from the `sequence` based predictions. The avalanche of new data sets can be harnessed by integrating different types of data and studying them computationally, by taking into account sequences, regulatory relationships, and kinetics simultaneously, in our effort to elucidate mechanisms and functions of miRNAs.

References

1. N. Rajewsky, MicroRNA target predictions in animals, *Nature Genetics.* **38**, S8–13, (2006).
2. F. Xiao, Z. Zuo, G. Cai, S. Kang, X. Gao, and T. Li, mirecords: an integrated resource for microrna-target interactions, *Nucleic Acids Res.* (2008).

3. E. Levine, T. Kuhlman, Z. Zhang, and T. Hwa, Quantitative characteristics of gene regulation by small rna, *PLoS Biology.* **5**(9), e229, (2007a).
4. Y. Shimoni, G. Friedlander, G. Hetzroni, G. Niv, S. Altuvia, O. Biham, and H. Margalit, Regulation of gene expression by small non-coding RNAs: a quantitative view, *Mol Syst Biol.* **7**(138), (2007).
5. E. Levine, P. McHale, and H. Levine, Small regulatory rnas may sharpen spatial expression patterns, *PLoS Comput Biol.* **3**(11), e233, (2007c).
6. T. Du and P. Zamore, microprimer: the biogenesis and function of microrna, *Development.* **132**(21), 4645–52, (2005).
7. E. Levine, E. B. Jacob, and H. Levine, Target-specific and global effectors in gene regulation by microrna, *Biophys. J.* **93**(11), L52–4, (2007b).
8. W. Filipowicz, S. Bhattacharyya, and N. Sonenberg, Mechanisms of post-transcriptional regulation by micrornas: are the answers in sight?, *Nat Rev Genet.* **9**(2), 102–14, (2008).
9. A. Krek, D. Grün, M. Poy, R. Wolf, L. Rosenberg, E. Epstein, P. MacMenamin, I. da Piedade, K. Gunsalus, M. Stoffel, and N. Rajewsky, Combinatorial microRNA target predictions, *Nature Genetics.* **37**(5), 495–500, (2005).
10. L. Lim, N. Lau, P. Garrett-Engele, A. Grimson, J. Schelter, J. Castle, D. Bartel, P. Linsley, and J. Johnson, Microarray analysis shows that some microRNAs down-regulate large numbers of target mRNAs, *Nature.* **433**(7027), 769–73, (2005).
11. D. Baek, J. Villen, C. Shin, F. Camargo, S. Gygi, and D. B. DP, The impact of micrornas on protein output, *Nature.* **455**(7209), 64–71, (2008).
12. M. Selbach, B. Schwanhaeusser, N. Thierfelder, Z. Fang, R. Khanin, and N. Rajewsky, Widespread changes in protein synthesis induced by micrornas, *Nature.* **455**(7209), 58–63, (2008).
13. J. Krützfeldt, N. Rajewsky, R. Braich, K. Rajeev, T. Tuschl, M. Manoharan, and M. Stoffel, Silencing of microRNAs in vivo with 'antagomirs', *Nature.* **438**(7068), 68509, (2005).
14. R. Friedman, K. Farh, C. Burge, and D. Bartel, Most mammalian mrnas are conserved targets of micrornas., *Genome Res.* **19**(1), 92–105, (2009).
15. J. Doench and P. Sharp, Specificity of microrna target selection in translational repression, *Genes Dev.* **16**(1), 504–11, (2004).
16. P. Maroney, Y. Yu, J. Fisher, and T. Nilsen, Evidence that micrornas are associated with translating messenger rnas in human cells, *Nat Struct Mol Biol.* **12**, 1102–7, (2006).
17. S. Bhattacharyya, R. Habermacher, U. Martine, E. Closs, and W. Filipowicz, Relief of microrna-mediated translational repression in human cells subjected to stress., *Cell.* **125**(6), 1111–1124, (2006).
18. A. Leung and P. Sharp, micrornas: A safeguard against turmoil?, *Cell.* **130**(4), 581–585, (2007).
19. H. Bolouri and E. Davidson, Modelling transcriptional regulatory networks, *BioEssays.* **24**, 1118–1129, (2002).
20. V. Ambros, The functions of animal microRNAs, *Nature.* **431**(7006), 350–5, (2004).
21. R. Jackson and N. Standart, How do micrornas regulate gene expression?, *Sci. STKE.* **367**, (2007).
22. R. Khanin and V. Vinciotti, Computational modeling of post-transcriptional gene regulation by micrornas, *J. Comput. Biol.* **15**(3), 305–316, (2008).
23. A. Grimson, K. Farh, W. Johnston, P. Garrett-Engele, L. Lim, and D.P.Bartel, Microrna targeting specificity in mammals: Determinants beyond seed pairing, *Mol. Cell.* **27**(1), 91–105, (2007).

24. R. Khanin and D. Higham, A multi-step model for microrna-mediated gene regulation, *Biosystems.* (2009).
25. O. Hobert, Gene regulation by transcription factors and micrornas, *Science.* **319**, 1785–1786, (2008).
26. R. Shalgi, D. Lieber, M. Oren, and Y. Pilpel, Global and local architecture of the mammalian microrna-transcription factor regulatory network, *PLoS Comput. Biol.* **3**(7), e131, (2007).
27. A. Tuccoli, L. Poliseno, and G. Rainaldi, mirnas regulate mirnas: coordinated transcriptional and post-transcriptional regulation, *Cell Cycle.* **5**(21), (2006).
28. X. Wang and X. Wang, Systematic identification of microRNA functions by combining target prediction and expression profiling, *Nucleic Acids Research.* **34**(5), 1646–52, (2006).
29. S. Shen-Orr, R. Milo, S. Mangan, and U. Alon, Network motifs in the transcriptional regulation network of Escherichia coli, *Nature Genetics.* **31**(1), 64–68, (2002).
30. R. Khanin, V. Vinciotti, and E. Wit, Reconstructing repressor protein levels from expression of gene targets in Escherichia coli, *PNAS.* **103**(49), 18592–6, (2006).
31. E. Yang, E. van Nimwegen, M. Zavolan, N. Rajewsky, M. Schroeder, M. Magnasco, and J. J. Darnell, Decay rates of human mrnas: correlation with functional characteristics and sequence attributes, *Genome Res.* **13**(8), 1863–1872, (2003).
32. S. Rogers, R. Khanin, and M. Girolami, Bayesian model-based inference of transcription factor activity, *BMC Bioinformatics.* **8**(Suppl 2), S2, (2007).
33. S. Lall, D. Grn, A. Krek, K. Chen, Y. Wang, C. D. C. P. Sood, T. Colombo, N. Bray, P. Macmenamin, H. Kao, K. Gunsalus, L. Pachter, F. Piano, and N. Rajewsky, A genome-wide map of conserved microrna targets in c. elegans, *Curr Biol.* **16**(5), 460–471, (2006).
34. Q. Cui, Z. Yu, Y. Pan, E. Purisima, and E. Wang, Micrornas preferentially target the genes with high transcriptional regulation complexity, *Biochem Biophys Res Commun.* **352**(3), 733–8, (2007).
35. X. Yu, J. Lin, D. Zack, J. Mendell, and J. Qian, Analysis of regulatory network topology reveals functionally distinct classes of micrornas, *Nucleic Acids Res.* **36**(20), 6494–503, (2008).
36. Q. Cui, Z. Yu, E. Purisima, and E. Wang, Principles of microRNA regulation of a human cellular signaling network, *Mol. Syst. Biol.* **2**, 46, (2006).
37. H. Liang and W. Li, Microrna regulation of human protein protein interaction network, *RNA.* **13**(9), 1402–8, (2007).
38. C. Hsu, H. Juan, and H. Huang, Characterization of microrna-regulated protein-protein interaction network, *Proteomics.* **8**, 1975–9, (2008).
39. E. Hornstein and N. Shomron, Canalization of development by microRNAs, *Nature Genet.* **38**(S20-4), 462–8, (2006).
40. O. Hobert, Common logic of transcription factor and microRNA action, *Trends Biochem Sci.* **29**(9), 462–8, (2004).
41. N. Martinez, M. Ow, M. Barrasa, M. Hammell, R. Sequerra, L. Doucette-Stamm, F. Roth, V. Ambros, and A. W. AJ, A c. elegans genome-scale microrna network contains composite feedback motifs with high flux capacity, *Genes Dev.* **22**(18), 2535–49, (2008).
42. B. Aguda, Y. Kim, M. Piper-Hunter, A. Friedman, and C. Marsh, Microrna regulation of a cancer network: Consequences of the feedback loops involving mir-17-92, e2f, and myc, *Proc Natl Acad Sci USA.* **105**(50), 19678–83, (2008).
43. A. Sinha, V. Kaimal, J. Chen, and A. Jegga, Dissecting microregulation of a master regulatory network, *BMC Genomics.* **9**(88), (2008).

44. J. Tsang, J. Zhu, and A. van Oudenaarden, Microrna-mediated feedback and feed-forward loops are recurrent network motifs in mammals, *Mol Cell.* **26**(5), 753–67, (2007).
45. V. Zhdanov, Bistability in gene transcription: interplay of messenger rna, protein, and nonprotein coding rna, *Biosystems.* **95**(1), 75–81, (2009).
46. P. Sood, A. Krek, M. Zavolan, G. Macino, and N. Rajewsky, Cell-type-specific signatures of microRNAs on target mRNA expression, *PNAS.* **103**(8), 2746–51, (2006).
47. Z. Xie, H. Yang, W. Liu, and M. Hwang, The role of microrna in the delayed negative feedback regulation of gene expression, *Biochem Biophys Res Commun.* **358**(3), 722–726, (2007).
48. T. Nissan and R. Parker, Computational analysis of mirna-mediated repression of translation: implications for models of translation initiation inhibition, *RNA.* **14**(8), 1480–91, (2008).
49. M. Marba, Stochastic modelling of post-transcriptional gene regulation by mirnas. (2009).

GLOSSARY OF GENOMICS AND BIOINFORMATICS TERMS*

A

Accession number (GenBank)

The accession number is the unique identifier assigned to the entire sequence record when the record is submitted to GenBank. The GenBank accession number is a combination of letters and numbers that are usually in the format of one letter followed by five digits (e.g., M12345) or two letters followed by six digits (e.g., AC123456). The accession number for a particular record will not change even if the author submits a request to change some of the information in the record. Take note that an accession number is a unique identifier for a complete sequence record, while a Sequence Identifier, such as a Version, GI, or ProteinID, is an identification number assigned just to the sequence data. The NCBI Entrez System is searchable by accession number using the Accession [ACCN] search field.

Accession number (RefSeq)

This accession number is the unique identification number for a complete RefSeq sequence record. RefSeq accession numbers are written in the following format: two letters followed by an underscore and six digits (e.g., NT_123456). The first two letters of the RefSeq accession number indicate the type of sequence included in the record as described below:

* NT_123456 constructed genomic contigs
* NM_123456 mRNAs (actually the cDNA sequences constructed from mRNA)
* NP_123456 proteins
* NC_123456 chromosomes

Adenine (A)
   A nitrogenous base, one member of the base pair AT (adenine-thymine).
   See also: base pair, nucleotide

---

*This glossary was partially compiled based on publicly available glossary of genomics terms at the Human Genome website: http://www.ornl.gov/sci/techresources/Human_Genome

Allele
Alternative form of a genetic locus; a single allele for each locus is inherited from each parent (e.g., at a locus for eye color the allele might result in blue or brown eyes).

Alternative splicing
Different ways of combining a gene's exons to make variants of the complete protein.

Amino acid
Any of a class of 20 molecules that are combined to form proteins in living things. The sequence of amino acids in a protein and hence protein function are determined by the genetic code.

Amplification
An increase in the number of copies of a specific DNA fragment; can be in vivo or in vitro.
See also: cloning, polymerase chain reaction

Annotation
Adding pertinent information such as gene coded for, amino acid sequence, or other commentary to the database entry of raw sequence of DNA bases.

Antisense
Nucleic acid that has a sequence exactly opposite to an mRNA molecule; binds to the mRNA molecule to prevent a protein from being made.

Apoptosis
Programmed cell death, the body's normal method of disposing of damaged, unwanted, or unneeded cells.

Arrayed library
Individual primary recombinant clones (hosted in phage, cosmid, YAC, or other vector) that are placed in two-dimensional arrays in microtiter dishes. Each primary clone can be identified by the identity of the plate and the clone location (row and column) on that plate. Arrayed libraries of clones can be used for many applications, including screening for a specific gene or genomic region of interest.

Assembly
Putting sequenced fragments of DNA into their correct chromosomal positions.

Autosome
A chromosome not involved in sex determination. The diploid human genome consists of a total of 46 chromosomes: 22 pairs of autosomes, and 1 pair of sex chromosomes (the X and Y chromosomes).

B

Bacterial artificial chromosome (BAC)
A vector used to clone DNA fragments (100- to 300-kb insert size; average, 150 kb) in Escherichia coli cells. Based on naturally occurring F-factor plasmid found in the bacterium E. coli.

Base
One of the molecules that form DNA and RNA molecules.

Base pair (bp)
Two nitrogenous bases (adenine and thymine or guanine and cytosine) held together by weak bonds. Two strands of DNA are held together in the shape of a double helix by the bonds between base pairs.

Base sequence
The order of nucleotide bases in a DNA molecule; determines structure of proteins encoded by that DNA.

Base sequence analysis
A method, sometimes automated, for determining the base sequence.

Bioinformatics
The science of managing and analyzing biological data using advanced computing techniques. Especially important in analyzing genomic and transciptomics research data.

BLAST
A computer program that identifies homologous (similar) genes in different organisms, such as human, fruit fly, or nematode.

C

Cancer
Diseases in which abnormal cells divide and grow unchecked. Cancer can spread from its original site to other parts of the body and can be fatal.

Candidate gene
A gene located in a chromosome region suspected of being involved in a disease.

Capillary array
Gel-filled silica capillaries used to separate fragments for DNA sequencing. The small diameter of the capillaries permit the application of higher electric fields, providing high speed, high throughput separations that are significantly faster than traditional slab gels.

Carcinogen
Something which causes cancer to occur by causing changes in a cell's DNA.

cDNA or complementary DNA
DNA that is synthesized in the laboratory from a messenger RNA template

cDNA library
A collection of DNA sequences that code for genes. The sequences are generated in the laboratory from mRNA sequences.

CDS
The coding sequence or the portion of a nucleotide sequence that makes up the triplet codons that actually code for amino acids.

Cell
The basic unit of any living organism that carries on the biochemical processes of life.

Chromosomal deletion
The loss of part of a chromosome's DNA.

Chromosomal inversion
Chromosome segments that have been turned 180 degrees. The gene sequence for the segment is reversed with respect to the rest of the chromosome.

Chromosome
The self-replicating genetic structure of cells containing the cellular DNA that bears in its nucleotide sequence the linear array of genes. In prokaryotes, chromosomal DNA is circular, and the entire genome is carried on one chromosome. Eukaryotic genomes consist of a number of chromosomes whose DNA is associated with different kinds of proteins.

Chromosome region p
A designation for the short arm of a chromosome.

Chromosome region q
A designation for the long arm of a chromosome.

Clone
An exact copy made of biological material such as a DNA segment (e.g., a gene or other region), a whole cell, or a complete organism.

Clone bank
See: genomic library

Cloning
Using specialized DNA technology to produce multiple, exact copies of a single gene or other segment of DNA to obtain enough material for further study. This process, used by researchers in the Human Genome Project, is referred to as cloning DNA. The resulting cloned (copied) collections of DNA molecules are called clone libraries. A second type of cloning exploits the natural process of cell division to make many copies of an entire cell. The genetic makeup of these cloned cells, called a cell line, is identical to the original cell. A third type of cloning produces complete, genetically identical animals such as the famous Scottish sheep, Dolly.

Cloning vector
DNA molecule originating from a virus, a plasmid, or the cell of a higher organism into which another DNA fragment of appropriate size can be integrated without loss of the vector's capacity for self-replication; vectors introduce foreign

DNA into host cells, where the DNA can be reproduced in large quantities. Examples are plasmids, cosmids, and yeast artificial chromosomes; vectors are often recombinant molecules containing DNA sequences from several sources.

Codon
See: genetic code

Comparative genomics
The study of human genetics by comparisons with model organisms such as mice, the fruit fly, and the bacterium E. coli.

Complementary DNA (cDNA)
DNA that is synthesized in the laboratory from a messenger RNA template.

Complementary sequence
Nucleic acid base sequence that can form a double-stranded structure with another DNA fragment by following base-pairing rules (A pairs with T and C with G). The complementary sequence to GTAC for example, is CATG.

Computational biology
See: bioinformatics

Conserved sequence
A base sequence in a DNA molecule (or an amino acid sequence in a protein) that has remained essentially unchanged throughout evolution.

Constitutive ablation
Gene expression that results in cell death.

Contig
Group of cloned (copied) pieces of DNA representing overlapping regions of a particular chromosome.

Contig map
A map depicting the relative order of a linked library of overlapping clones representing a complete chromosomal segment.

Cosmid

Artificially constructed cloning vector containing the cos gene of phage lambda. Cosmids can be packaged in lambda phage particles for infection into E. coli; this permits cloning of larger DNA fragments (up to 45kb) than can be introduced into bacterial hosts in plasmid vectors.

Crossing over

The breaking during meiosis of one maternal and one paternal chromosome, the exchange of corresponding sections of DNA, and the rejoining of the chromosomes. This process can result in an exchange of alleles between chromosomes.

Cytogenetics

The study of the physical appearance of chromosomes.

Cytological band

An area of the chromosome that stains differently from areas around it.

Cytological map

A type of chromosome map whereby genes are located on the basis of cytological findings obtained with the aid of chromosome mutations.

Cytosine (C)

A nitrogenous base, one member of the base pair GC (guanine and cytosine) in DNA.

D

Deletion

A loss of part of the DNA from a chromosome; can lead to a disease or abnormality.

Deletion map

A description of a specific chromosome that uses defined mutations --specific deleted areas in the genome-- as 'biochemical signposts,' or markers for specific areas.

Deoxyribonucleotide

See: nucleotide

Deoxyribose
A type of sugar that is one component of DNA (deoxyribonucleic acid).

Diploid
A full set of genetic material consisting of paired chromosomes, one from each parental set. Most animal cells except the gametes have a diploid set of chromosomes. The diploid human genome has 46 chromosomes.

Directed mutagenesis
Alteration of DNA at a specific site and its reinsertion into an organism to study any effects of the change.

Directed sequencing
Successively sequencing DNA from adjacent stretches of chromosome.

Disease-associated genes
Alleles carrying particular DNA sequences associated with the presence of disease.

DNA (deoxyribonucleic acid)
The molecule that encodes genetic information. DNA is a double-stranded molecule held together by weak bonds between base pairs of nucleotides. The four nucleotides in DNA contain the bases adenine (A), guanine (G), cytosine (C), and thymine (T). In nature, base pairs form only between A and T and between G and C; thus the base sequence of each single strand can be deduced from that of its partner.

DNA probe
See: probe

DNA repair genes
Genes encoding proteins that correct errors in DNA sequencing.

DNA replication
The use of existing DNA as a template for the synthesis of new DNA strands. In humans and other eukaryotes, replication occurs in the cell nucleus.

DNA sequence
The relative order of base pairs, whether in a DNA fragment, gene, chromosome, or an entire genome.
See also: base sequence analysis

Domain
A discrete portion of a protein with its own function. The combination of domains in a single protein determines its overall function.

Double helix
The twisted-ladder shape that two linear strands of DNA assume when complementary nucleotides on opposing strands bond together.

E

Electrophoresis
A method of separating large molecules (such as DNA fragments or proteins) from a mixture of similar molecules. An electric current is passed through a medium containing the mixture, and each kind of molecule travels through the medium at a different rate, depending on its electrical charge and size. Agarose and acrylamide gels are the media commonly used for electrophoresis of proteins and nucleic acids.

Electroporation
A process using high-voltage current to make cell membranes permeable to allow the introduction of new DNA; commonly used in recombinant DNA technology.

Embryonic stem (ES) cells
An embryonic cell that can replicate indefinitely, transform into other types of cells, and serve as a continuous source of new cells.

Endonuclease
See: restriction enzyme

Enzyme
A protein that acts as a catalyst, speeding the rate at which a biochemical reaction proceeds but not altering the direction or nature of the reaction.

Epistasis
One gene interferes with or prevents the expression of another gene located at a different locus.

Escherichia coli
Common bacterium that has been studied intensively by geneticists because of its small genome size, normal lack of pathogenicity, and ease of growth in the laboratory.

Eukaryote
Cell or organism with membrane-bound, structurally discrete nucleus and other well-developed subcellular compartments. Eukaryotes include all organisms except viruses, bacteria, and bluegreen algae.

Evolutionarily conserved sequence
A base sequence in a DNA molecule (or an amino acid sequence in a protein) that has remained essentially unchanged throughout evolution

Exogenous DNA
DNA originating outside an organism that has been introduced into the organism.

Exon
The protein-coding DNA sequence of a gene.
See also: intron

Exonuclease
An enzyme that cleaves nucleotides sequentially from free ends of a linear nucleic acid substrate.

Expressed gene
See: gene expression

Expressed sequence tag (EST)
A short strand of DNA that is a part of a cDNA molecule and can act as identifier of a gene. Used in locating and mapping genes.

F

Fluorescence in situ hybridization (FISH)
A physical mapping approach that uses fluorescein tags to detect hybridization of probes with metaphase chromosomes and with the less-condensed somatic interphase chromatin.

Full gene sequence
The complete order of bases in a gene. This order determines which protein a gene will produce.

Functional genomics
The study of genes, their resulting proteins, and the role played by the proteins in the body's biochemical processes.

G

Gamete
Mature male or female reproductive cell (sperm or ovum) with a haploid set of chromosomes (23 for humans).

GC-rich area
Many DNA sequences carry long stretches of repeated G and C which often indicate a gene-rich region.

Gel electrophoresis
See: electrophoresis

Gene
The fundamental physical and functional unit of heredity. A gene is an ordered sequence of nucleotides located in a particular position on a particular chromosome that encodes a specific functional product (i.e., a protein or RNA molecule).
See also: gene expression

Gene amplification
Repeated copying of a piece of DNA; a characteristic of tumor cells.

Gene chip technology
Development of cDNA microarrays from a large number of genes. Used to monitor and measure changes in gene expression for each gene represented on the chip.

Gene expression
The process by which a gene's coded information is converted into the structures present and operating in the cell. Expressed genes include those that are transcribed into mRNA and then translated into protein and those that are transcribed into RNA but not translated into protein (e.g., transfer and ribosomal RNAs).

Gene family
Group of closely related genes that make similar products.

Gene library
See: genomic library

Gene locus (pl. loci)
Gene's position on a chromosome or other chromosome marker; also, the DNA at that position. The use of locus is sometimes restricted to mean expressed DNA regions

Gene mapping
Determination of the relative positions of genes on a DNA molecule (chromosome or plasmid) and of the distance, in linkage units or physical units, between them.

Gene name
Official name assigned to a gene. According to the Guidelines for Human Gene Nomenclature developed by the HUGO Gene Nomenclature Committee, it should be brief and describe the function of the gene.

Gene ontology
A controlled vocabulary of terms relating to molecular function, biological process, or cellular components developed by the Gene Ontology Consortium. A controlled vocabulary allows scientists to use consistent terminology when describing the roles of genes and proteins in cells.

Gene pool
All the variations of genes in a species.

Gene prediction
Predictions of possible genes made by a computer program based on how well a stretch of DNA sequence matches known gene sequences

Gene product
The biochemical material, either RNA or protein, resulting from expression of a gene. The amount of gene product is used to measure how active a gene is; abnormal amounts can be correlated with disease-causing alleles.

Gene testing
See: genetic testing, genetic screening

Gene therapy
An experimental procedure aimed at replacing, manipulating, or supplementing nonfunctional or misfunctioning genes with healthy genes.

Gene transfer
Incorporation of new DNA into and organism's cells, usually by a vector such as a modified virus. Used in gene therapy.

Genetic code
The sequence of nucleotides, coded in triplets (codons) along the mRNA, that determines the sequence of amino acids in protein synthesis. A gene's DNA sequence can be used to predict the mRNA sequence, and the genetic code can in turn be used to predict the amino acid sequence.

Genetic engineering
Altering the genetic material of cells or organisms to enable them to make new substances or perform new functions.

Genetic engineering technology
See: recombinant DNA technology

Genetic illness
Sickness, physical disability, or other disorder resulting from the inheritance of one or more deleterious alleles.

Genetic map
See: linkage map

Genetic marker
A gene or other identifiable portion of DNA whose inheritance can be followed.

Genetic material
See: genome

Genetic polymorphism
Difference in DNA sequence among individuals, groups, or populations (e.g., genes for blue eyes versus brown eyes).

Genetic predisposition
Susceptibility to a genetic disease. May or may not result in actual development of the disease.

Genetic screening
Testing a group of people to identify individuals at high risk of having or passing on a specific genetic disorder.

Gene symbol
Symbols for human genes are usually designated by scientists who discover the genes. The symbols are created using the Guidelines for Human Gene Nomenclature developed by the HUGO Gene Nomenclature Committee. Gene symbols usually consist of no more than six upper case letters or combination of uppercase letters and Arabic numbers. Gene symbols should start with the first letters of the gene name. For example, the gene symbol for insulin is "INS." A gene symbol must be submitted to HUGO for approval before it can be considered an official gene symbol.

Genetic testing
Analyzing an individual's genetic material to determine predisposition to a particular health condition or to confirm a diagnosis of genetic disease.

Genetics
The study of inheritance patterns of specific traits.

Genome
All the genetic material in the chromosomes of a particular organism; its size is generally given as its total number of base pairs.

Genomic library
A collection of clones made from a set of randomly generated overlapping DNA fragments that represent the entire genome of an organism.
See also: library, arrayed library

Genomic sequence
See: DNA

Genomics
The study of genes and their function.

Genotype
The genetic constitution of an organism, as distinguished from its physical appearance (its phenotype).

Germ cell
Sperm and egg cells and their precursors. Germ cells are haploid and have only one set of chromosomes (23 in all), while all other cells have two copies (46 in all).

Germ line
The continuation of a set of genetic information from one generation to the next.

Germ line gene therapy
An experimental process of inserting genes into germ cells or fertilized eggs to cause a genetic change that can be passed on to offspring. May be used to alleviate effects associated with a genetic disease.

Germ line genetic mutation
See: mutation

Guanine (G)
A nitrogenous base, one member of the base pair GC (guanine and cytosine) in DNA.
See also: base pair, nucleotide

H

Haploid
A single set of chromosomes (half the full set of genetic material) present in the egg and sperm cells of animals and in the egg and pollen cells of plants. Human beings have 23 chromosomes in their reproductive cells.
See also: diploid

Haplotype
A way of denoting the collective genotype of a number of closely linked loci on a chromosome.

Hemizygous
Having only one copy of a particular gene. For example, in humans, males are hemizygous for genes found on the Y chromosome.

Hereditary cancer
Cancer that occurs due to the inheritance of an altered gene within a family.
See also: sporadic cancer

Heterozygosity
The presence of different alleles at one or more loci on homologous chromosomes.

Heterozygote
See: heterozygosity

Highly conserved sequence
DNA sequence that is very similar across several different types of organisms.

High-throughput sequencing
A fast method of determining the order of bases in DNA.

Homeobox
A short stretch of nucleotides whose base sequence is virtually identical in all the genes that contain it. Homeoboxes have been found in many organisms from fruit flies to human beings. In the fruit fly, a homeobox appears to determine when particular groups of genes are expressed during development.

Homolog
A member of a chromosome pair in diploid organisms or a gene that has the same origin and functions in two or more species.

Homologous chromosome
Chromosome containing the same linear gene sequences as another, each derived from one parent.

Homologous recombination
Swapping of DNA fragments between paired chromosomes.

Homology
Similarity in DNA or protein sequences between individuals of the same species or among different species.

Homozygote
An organism that has two identical alleles of a gene.
See also: heterozygote

Homozygous
See: homozygote

Human artificial chromosome (HAC)
A vector used to hold large DNA fragments.
See also: chromosome, DNA

Human gene therapy
See: gene therapy

Hybridization
The process of joining two complementary strands of DNA or one each of DNA and RNA to form a double-stranded molecule.

I

Immunotherapy
Using the immune system to treat disease, for example, in the development of vaccines. May also refer to the therapy of diseases caused by the immune system.

Imprinting

A phenomenon in which the disease phenotype depends on which parent passed on the disease gene. For instance, both Prader-Willi and Angelman syndromes are inherited when the same part of chromosome 15 is missing. When the father's complement of 15 is missing, the child has Prader-Willi, but when the mother's complement of 15 is missing, the child has Angelman syndrome.

In situ hybridization

Use of a DNA or RNA probe to detect the presence of the complementary DNA sequence in cloned bacterial or cultured eukaryotic cells.

In vitro

Studies performed outside a living organism such as in a laboratory.

In vivo

Studies carried out in living organisms.

Inherit

In genetics, to receive genetic material from parents through biological processes.

Insertion

A chromosome abnormality in which a piece of DNA is incorporated into a gene and thereby disrupts the gene's normal function.
See also: chromosome, DNA, gene, mutation

Insertional mutation
See: insertion

Interphase

The period in the cell cycle when DNA is replicated in the nucleus; followed by mitosis.

Intron

DNA sequence that interrupts the protein-coding sequence of a gene; an intron is transcribed into RNA but is cut out of the message before it is translated into protein. Many introns are shown to contain small RNA-coding sequences
See also: exon

Isoenzyme
An enzyme performing the same function as another enzyme but having a different set of amino acids. The two enzymes may function at different speeds.

K

Karyotype
A photomicrograph of an individual's chromosomes arranged in a standard format showing the number, size, and shape of each chromosome type; used in low-resolution physical mapping to correlate gross chromosomal abnormalities with the characteristics of specific diseases.

Kilobase (kb)
Unit of length for DNA fragments equal to 1000 nucleotides.

Knockdown
Deactivation of specific genes via RNA interference mechanism; used in laboratory organisms to study gene function

Knockout
Deactivation of specific genes via specific DNA binding mechanisms; used in laboratory organisms to study gene function.

L

Library
An unordered collection of clones (i.e., cloned DNA from a particular organism) whose relationship to each other can be established by physical mapping.

Localize
Determination of the original position (locus) of a gene or other marker on a chromosome.

Locus (pl. loci)
The position on a chromosome of a gene or other chromosome marker; also, the DNA at that position. The use of locus is sometimes restricted to mean expressed DNA regions.

Long-range restriction mapping
Restriction enzymes are proteins that cut DNA at precise locations. Restriction maps depict the chromosomal positions of restriction-enzyme cutting sites. These are used as biochemical "signposts," or markers of specific areas along the chromosomes. The map will detail the positions where the DNA molecule is cut by particular restriction enzymes.

M

Macrorestriction map
Map depicting the order of and distance between sites at which restriction enzymes cleave chromosomes.

Mapping
See: gene mapping, linkage map, physical map

Marker
See: genetic marker

Mass spectrometry
An instrument used to identify chemicals in a substance by their mass and charge.

Megabase (Mb)
Unit of length for DNA fragments equal to 1 million nucleotides and roughly equal to 1 cM (centimorgan).

Meiosis
The process of two consecutive cell divisions in the diploid progenitors of sex cells. Meiosis results in four rather than two daughter cells, each with a haploid set of chromosomes.
See also: mitosis

Messenger RNA (mRNA)
RNA that serves as a template for protein synthesis.

Metaphase
A stage in mitosis or meiosis during which the chromosomes are aligned along the equatorial plane of the cell.

Microarray
Sets of miniaturized chemical reaction areas that may also be used to test DNA or RNA fragments, antibodies, or proteins.

Micronuclei
Chromosome fragments that are not incorporated into the nucleus at cell division.

microRNAs (miRNA)
single-stranded RNA molecules of 19-24 nucleotides in length, which regulate gene expression. miRNAs are encoded by genes from whose DNA they are transcribed but miRNAs are not translated into protein (non-coding RNA); instead each primary transcript (a pri-miRNA) is processed into a short stem-loop structure called a pre-miRNA and finally into a functional miRNA. Mature miRNA molecules are partially complementary to one or more messenger RNA (mRNA) molecules, and their main function is to down-regulate gene expression via mRNA degradation and/or translation inhibition. They were first described in 1993 by Lee and colleagues in the Victor Ambros lab, yet the term microRNA was only introduced in 2001 in a set of three articles in *Science*.

Mitochondrial DNA
The genetic material found in mitochondria, the organelles that generate energy for the cell. Not inherited in the same fashion as nucleic DNA.

Mitosis
The process of nuclear division in cells that produces daughter cells that are genetically identical to each other and to the parent cell.
See also: meiosis

Model organisms
A laboratory animal or other organism useful for research.

Modeling
The use of statistical analysis, mathematical equations, computer simulation.

Monogenic disorder
A disorder caused by mutation of a single gene.

Monogenic inheritance
See: monogenic disorder

Monosomy
Possessing only one copy of a particular chromosome instead of the normal two copies.

Mouse model
See: model organisms

Multiplexing
A laboratory approach that performs multiple sets of reactions in parallel (simultaneously); greatly increasing speed and throughput.

Murine
Organism in the genus Mus. A rat or mouse.

Mutagen
An agent that causes a permanent genetic change in a cell. Does not include changes occurring during normal genetic recombination.

Mutagenicity
The capacity of a chemical or physical agent to cause permanent genetic alterations.
See also: somatic cell genetic mutation

Mutation
Any heritable change in DNA sequence.
See also: polymorphism

N

Nitrogenous base
A nitrogen-containing molecule having the chemical properties of a base. DNA contains the nitrogenous bases adenine (A), guanine (G), cytosine (C), and thymine (T).

Non-coding RNA (ncRNA)
ncRNA is transcribed from DNA, but not translated into protein. It often plays a regulatory role. Some probably most ancient ncRNAs encompass catalytic functions and are called ribozymes

Northern blot
A gel-based laboratory procedure that locates mRNA sequences on a gel that are complementary to a piece of DNA used as a probe.

Nuclear transfer
A laboratory procedure in which a cell's nucleus is removed and placed into an oocyte with its own nucleus removed so the genetic information from the donor nucleus controls the resulting cell. Such cells can be induced to form embryos. This process was used to create the cloned sheep "Dolly".

Nucleic acid
A large molecule composed of nucleotide subunits.
See also: DNA

Nucleolar organizing region
A part of the chromosome containing rRNA genes.

Nucleotide
A subunit of DNA or RNA consisting of a nitrogenous base (adenine, guanine, thymine, or cytosine in DNA; adenine, guanine, uracil, or cytosine in RNA), a phosphate molecule, and a sugar molecule (deoxyribose in DNA and ribose in RNA). Thousands of nucleotides are linked to form a DNA or RNA molecule.

Nucleus
The cellular organelle in eukaryotes that contains most of the genetic material.

O

Oligo
See: oligonucleotide

Oligonucleotide
A molecule usually composed of 25 or fewer nucleotides; used as a DNA synthesis primer.
See also: nucleotide

Oncogene

A gene, one or more forms of which is associated with cancer. Many oncogenes are involved, directly or indirectly, in controlling the rate of cell growth.

Open reading frame (ORF)

The sequence of DNA or RNA located between the start-code sequence (initiation codon) and the stop-code sequence (termination codon).

Operon

A set of genes transcribed under the control of an operator gene.

Overlapping clones
See: genomic library

P

P1-derived artificial chromosome (PAC)

One type of vector used to clone DNA fragments (100- to 300-kb insert size; average, 150 kb) in Escherichia coli cells. Based on bacteriophage (a virus) P1 genome.
See also: cloning vector

Peptide
Two or more amino acids joined by a bond called a "peptide bond".
See also: polypeptide

Phage
A virus for which the natural host is a bacterial cell.

Pharmacogenomics

The study of the interaction of an individual's genetic makeup and response to a drug.

Phenotype

The physical characteristics of an organism or the presence of a disease that may or may not be genetic.
See also: genotype

Physical map

A map of the locations of identifiable landmarks on DNA (e.g., restriction-enzyme cutting sites, genes), regardless of inheritance. Distance is measured in base pairs. For the human genome, the lowest-resolution physical map is the banding patterns on the 24 different chromosomes; the highest-resolution map is the complete nucleotide sequence of the chromosomes.

Plasmid

Autonomously replicating extra-chromosomal circular DNA molecules, distinct from the normal bacterial genome and nonessential for cell survival under nonselective conditions. Some plasmids are capable of integrating into the host genome. A number of artificially constructed plasmids are used as cloning vectors.

Pleiotropy

One gene that causes many different physical traits such as multiple disease symptoms.

Pluripotency

The potential of a cell to develop into more than one type of mature cell, depending on environment.

Polymerase chain reaction (PCR)

A method for amplifying a DNA base sequence using a heat-stable polymerase and two 20-base primers, one complementary to the (+) strand at one end of the sequence to be amplified and one complementary to the (-) strand at the other end. Because the newly synthesized DNA strands can subsequently serve as additional templates for the same primer sequences, successive rounds of primer annealing, strand elongation, and dissociation produce rapid and highly specific amplification of the desired sequence. PCR also can be used to detect the existence of the defined sequence in a DNA sample.

Polymerase, DNA or RNA

Enzyme that catalyzes the synthesis of nucleic acids on preexisting nucleic acid templates, assembling RNA from ribonucleotides or DNA from deoxyribonucleotides.

Polymorphism
Difference in DNA sequence among individuals that may underlie differences in health. Genetic variations occurring in more than 1% of a population would be considered useful polymorphisms for genetic linkage analysis.
See also: mutation

Polypeptide
A protein or part of a protein made of a chain of amino acids joined by a peptide bond.

Population genetics
The study of variation in genes among a group of individuals.

Positional cloning
A technique used to identify genes, usually those that are associated with diseases, based on their location on a chromosome.

Premature chromosome condensation (PCC)
A method of studying chromosomes in the interphase stage of the cell cycle.

Primer
Short preexisting polynucleotide chain to which new deoxyribonucleotides can be added by DNA polymerase.

Probe
Single-stranded DNA or RNA molecules of specific base sequence, labeled either radioactively or immunologically, that are used to detect the complementary base sequence by hybridization.

Prokaryote
Cell or organism lacking a membrane-bound, structurally discrete nucleus and other subcellular compartments. Bacteria are examples of prokaryotes.
See also: chromosome, eukaryote

Promoter
A DNA site to which RNA polymerase will bind and initiate transcription.

Pronucleus
The nucleus of a sperm or egg prior to fertilization.
See also: nucleus, transgenic

Protein

A large molecule composed of one or more chains of amino acids in a specific order; the order is determined by the base sequence of nucleotides in the gene that codes for the protein. Proteins are required for the structure, function, and regulation of the body's cells, tissues, and organs; and each protein has unique functions. Examples are hormones, enzymes, and antibodies.

Proteome

Proteins expressed by a cell or organ at a particular time and under specific conditions.

Proteomics

The study of the full set of proteins encoded by a genome.

Pseudogene

A sequence of DNA similar to a gene but nonfunctional; probably the remnant of a once-functional gene that accumulated mutations.

Purine

A nitrogen-containing, double-ring, basic compound that occurs in nucleic acids. The purines in DNA and RNA are adenine and guanine.
See also: base pair

Pyrimidine

A nitrogen-containing, single-ring, basic compound that occurs in nucleic acids. The pyrimidines in DNA are cytosine and thymine; in RNA, cytosine and uracil.
See also: base pair

R

Radiation hybrid

A hybrid cell containing small fragments of irradiated human chromosomes. Maps of irradiation sites on chromosomes for the human, rat, mouse, and other genomes provide important markers, allowing the construction of very precise STS maps indispensable to studying multifactorial diseases.

Recessive gene
A gene which will be expressed only if there are 2 identical copies or, for a male, if one copy is present on the X chromosome.

Reciprocal translocation
When a pair of chromosomes exchange exactly the same length and area of DNA. Results in a shuffling of genes.

Recombinant clone
Clone containing recombinant DNA molecules.

Recombinant DNA molecules
A combination of DNA molecules of different origin that are joined using recombinant DNA technologies.

Recombinant DNA technology
Procedure used to join together DNA segments in a cell-free system (an environment outside a cell or organism). Under appropriate conditions, a recombinant DNA molecule can enter a cell and replicate there, either autonomously or after it has become integrated into a cellular chromosome.

Recombination
The process by which progeny derive a combination of genes different from that of either parent. In higher organisms, this can occur by crossing over.
See also: crossing over, mutation

Regulatory region or sequence
A DNA base sequence that controls gene expression.

Repetitive DNA
Sequences of varying lengths that occur in multiple copies in the genome; it represents much of the human genome.

Reporter gene
See: marker

Resolution
Degree of molecular detail on a physical map of DNA, ranging from low to high.

Restriction enzyme, endonuclease
A protein that recognizes specific, short nucleotide sequences and cuts DNA at those sites. Bacteria contain over 400 such enzymes that recognize and cut more than 100 different DNA sequences.

Restriction fragment length polymorphism (RFLP)
Variation between individuals in DNA fragment sizes cut by specific restriction enzymes; polymorphic sequences that result in RFLPs are used as markers on both physical maps and genetic linkage maps. RFLPs usually are caused by mutation at a cutting site.

Restriction-enzyme cutting site
A specific nucleotide sequence of DNA at which a particular restriction enzyme cuts the DNA. Some sites occur frequently in DNA (e.g., every several hundred base pairs); others much less frequently (rare-cutter; e.g., every 10,000 base pairs).

Retroviral infection
The presence of retroviral vectors, such as some viruses, which use their recombinant DNA to insert their genetic material into the chromosomes of the host's cells. The virus is then propogated by the host cell.

Reverse transcriptase
An enzyme used by retroviruses to form a complementary DNA sequence (cDNA) from their RNA. The resulting DNA is then inserted into the chromosome of the host cell.

Ribonucleotide
   See: nucleotide

Ribose
   The five-carbon sugar that serves as a component of RNA.
   See also: ribonucleic acid, deoxyribose

Ribosomal RNA (rRNA)
   A class of RNA found in the ribosomes of cells.

Ribosomes
Small cellular components composed of specialized ribosomal RNA and protein; site of protein synthesis.
See also: RNA

RNA (Ribonucleic acid)
A chemical found in the nucleus and cytoplasm of cells; it plays an important role in protein synthesis and other chemical activities of the cell. The structure of RNA is similar to that of DNA. There are several classes of RNA molecules, including messenger RNA, transfer RNA, ribosomal RNA, and other small RNAs, each serving a different purpose.

RNA interference (RNAi)
A gene-silencing process in which double-stranded RNAs trigger the destruction of specific RNAs.

S

Sanger sequencing
    A widely used method of determining the order of bases in DNA.
    See also: sequencing, shotgun sequencing

Satellite
A chromosomal segment that branches off from the rest of the chromosome but is still connected by a thin filament or stalk.

Scaffold
In genomic mapping, a series of contigs that are in the right order but not necessarily connected in one continuous stretch of sequence.

Segregation
The normal biological process whereby the two pieces of a chromosome pair are separated during meiosis and randomly distributed to the germ cells.

Sequence
    See: base sequence

Sequence assembly
A process whereby the order of multiple sequenced DNA fragments is determined.

Sequence tagged site (STS)
Short (200 to 500 base pairs) DNA sequence that has a single occurrence in the human genome and whose location and base sequence are known. Detectable by polymerase chain reaction, STSs are useful for localizing and orienting the mapping and sequence data reported from many different laboratories and serve as landmarks on the developing physical map of the human genome. Expressed sequence tags (ESTs) are STSs derived from cDNAs.

Sequencing
Determination of the order of nucleotides (base sequences) in a DNA or RNA molecule or the order of amino acids in a protein.

Sequencing technology
The instrumentation and procedures used to determine the order of nucleotides in DNA.

Sex chromosome
The X or Y chromosome in human beings that determines the sex of an individual. Females have two X chromosomes in diploid cells; males have an X and a Y chromosome. The sex chromosomes comprise the 23rd chromosome pair in a karyotype.
See also: autosome

Sex-linked
Traits or diseases associated with the X or Y chromosome; generally seen in males.

Shotgun method
Sequencing method that involves randomly sequenced cloned pieces of the genome, with no foreknowledge of where the piece originally came from. This can be contrasted with "directed" strategies, in which pieces of DNA from known chromosomal locations are sequenced. Because there are advantages to both strategies, researchers use both random (or shotgun) and directed strategies in combination to sequence the human genome.

Single nucleotide polymorphism (SNP)
DNA sequence variations that occur when a single nucleotide (A, T, C, or G) in the genome sequence is altered.

Somatic cell
Any cell in the body except gametes and their precursors.
See also: gamete

Somatic cell gene therapy
Incorporating new genetic material into cells for therapeutic purposes. The new genetic material cannot be passed to offspring.
See also: gene therapy

Somatic cell genetic mutation
A change in the genetic structure that is neither inherited nor passed to offspring. Also called acquired mutations.
See also: germ line genetic mutation

Southern blotting
Transfer by absorption of DNA fragments separated in electrophoretic gels to membrane filters for detection of specific base sequences by radio-labeled complementary probes.

Splice site
Location in the DNA sequence where RNA removes the noncoding areas to form a continuous gene transcript for translation into a protein.

Sporadic cancer
Cancer that occurs randomly and is not inherited from parents..

Stem cell
Undifferentiated, primitive cells in the bone marrow that have the ability both to multiply and to differentiate into specific blood cells.

Substitution
In genetics, a type of mutation due to replacement of one nucleotide in a DNA sequence by another nucleotide or replacement of one amino acid in a protein by another amino acid.
See also: mutation

Suppressor gene
A gene that can suppress the action of another gene.

Syndrome
The group or recognizable pattern of symptoms or abnormalities that indicate a particular trait or disease.

Syngeneic
Genetically identical members of the same species.

Synteny
Genes occurring in the same order on chromosomes of different species.
See also: linkage, conserved sequence

Systems biology
A field that seeks to study the relationships and interactions between various parts of a biological system (metabolic pathways, organelles, cells, and organisms) and to integrate this information to understand how biological systems function

T

Tandem repeat sequences
Multiple copies of the same base sequence on a chromosome; used as markers in physical mapping.

Targeted mutagenesis
Deliberate change in the genetic structure directed at a specific site on the chromosome. Used in research to determine the targeted region's function.

Technology transfer
The process of transferring scientific findings from research laboratories to the commercial sector.

Telomerase
The enzyme that directs the replication of telomeres.

Telomere
The end of a chromosome. This specialized structure is involved in the replication and stability of linear DNA molecules.

Teratogenic
Substances such as chemicals or radiation that cause abnormal development of a embryo.

Thymine (T)
A nitrogenous base, one member of the base pair AT (adenine-thymine).
See also: base pair, nucleotide

Toxicogenomics
The study of how genomes respond to environmental stressors or toxicants. Combines genome-wide mRNA expression profiling with protein expression patterns using bioinformatics to understand the role of gene-environment interactions in disease and dysfunction.

Transcription
The synthesis of an RNA copy from a sequence of DNA (a gene); the first step in gene expression.
See also: translation

Transcription factor
A protein that binds to regulatory regions and helps control gene expression.

Transcriptome
The full complement of activated genes, mRNAs, or transcripts in a particular tissue at a particular time

Transfection
The introduction of foreign DNA into a host cell.
See also: cloning vector, gene therapy

Transfer RNA (tRNA)
A class of RNA having structures with triplet nucleotide sequences that are complementary to the triplet nucleotide coding sequences of mRNA. The role of tRNAs in protein synthesis is to bond with amino acids and transfer them to the

ribosomes, where proteins are assembled according to the genetic code carried by mRNA.

Transformation
A process by which the genetic material carried by an individual cell is altered by incorporation of exogenous DNA into its genome.

Transgenic
An experimentally produced organism in which DNA has been artificially introduced and incorporated into the organism's germ line.

Translation
The process in which the genetic code carried by mRNA directs the synthesis of proteins from amino acids.
See also: transcription

Translocation
A mutation in which a large segment of one chromosome breaks off and attaches to another chromosome.
See also: mutation

Transposable element
A class of DNA sequences that can move from one chromosomal site to another.

Trisomy
Possessing three copies of a particular chromosome instead of the normal two copies.

U

Uracil
A nitrogenous base normally found in RNA but not DNA; uracil is capable of forming a base pair with adenine.
See also: base pair, nucleotide

V

Vector
See: cloning vector

Version (GenBank)

Similar to the Protein ID for protein sequences, the version is a nucleotide sequence identification number assigned to each GenBank sequence. The format for this sequence identifier is accession.version (e.g., M12345.1). Whenever the author of a particular sequence record changes the sequence data in any way (even if just a single nucleotide is altered), the version number will be increased by an increment of one, while the accession number base remains constant. For example, M12345.1 would become M12345.2. Each sequence change also results in the assignment of a new GI number [link to GI entry]. Whenever an individual searches an NCBI sequence database, only the most recent version of a record is retrieved. Use NCBI's Sequence Revision History page to view the different GI numbers, version numbers, or update dates associated with a particular GenBank record.

Virus

A noncellular biological entity that can reproduce only within a host cell. Viruses consist of nucleic acid covered by protein; some animal viruses are also surrounded by membrane. Inside the infected cell, the virus uses the synthetic capability of the host to produce progeny virus.

W

Western blot

A technique used to identify and locate proteins based on their ability to bind to specific antibodies.

See also: DNA, Northern blot, protein, RNA, Southern blotting

Wild type

The form of an organism that occurs most frequently in nature.

X

X chromosome

One of the two sex chromosomes, X and Y.

See also: Y chromosome, sex chromosome

Xenograft
Tissue or organs from an individual of one species transplanted into or grafted onto an organism of another species, genus, or family. A common example is the use of pig heart valves in humans.

Y

Y chromosome
    One of the two sex chromosomes, X and Y.
    See also: X chromosome, sex chromosome

Yeast artificial chromosome (YAC)
    Constructed from yeast DNA, it is a vector used to clone large DNA fragments.
    See also: cloning vector, cosmid

Z

Zinc-finger protein
    A secondary feature of some proteins containing a zinc atom; a DNA-binding protein.

INDEX

3'-end heterogeneity   33

AD and AI prostate cancer   114
Agilent miRNA microarrays   3
amplification/deletion   152
AND-gate   177
androgen receptor   128
androgen-deprivation therapy   113
anticancer drugs   208
apoptosis signaling pathway   208
apoptosis-associated tyrosine
    kinase   140
applied biosystems TaqMan qPCR   3
array hybridization   6
avoiding normalization   18

B cell leukemia   155
background correction   36
B-cell maturation   97
BCL2 gene   102
BIC1 locus   154
Bioanalyzer   15
BLAST   71
breast cancer   199
Burkitt's lymphoma   152

caspase-3   209
CD34+ cells   98
cell cycle   205
cell death   205
cell development   2
cell differentiation   2
cell morphology   205
cell regulation   2
cellular regulatory network   169
chromatin immunoprecipitation   170

chromosomal breakpoints   159
chromosomal translocations   152
colon cancer   199
combinatorial enrichment
    analysis   198
combinatorial regulation   177
computational simulations   219
cycle threshold (Ct)   9
cyclin-dependent kinase inhibitor   126

DAVID   193
DIANA-microT   172
differential expression   19
diffuse large B-cell lymphoma   89
direct end-labeling   3
disease-linked SNPs   115

E2F1   210
ectopic expression of microRNA   125
ectopically overexpressed
    miRNA   173
EMBOSS   58
empirical Bayes   46
Eprimer3   58
error bounds   5
erythropoiesis   98
experimental design   2

fine-tuning of gene expression   220
Fisher exact test   193
fold change   5
fold-change plots   12
fragile sites   113, 152

G1/S check point pathway   214
Gaussian distribution   4

255

GC content   9
gene ontology   193
gene-within-a-gene   139
genomic instability   151
Gleason histological grade   139
GO enrichment analysis   193
GO terms   179

Hematopoiesis   94
hepatocellular carcinomas   210
histone methyltransferase   126
host-antagonistic gene pool   144
human papilloma virus   152
hypergeometric distribution   193

IDN-6556   209
ingenuity pathway analysis   202
intronic miRNA   137
isomiRs   31

Jaspar   174

kiss of the miRNA hypothesis   143

linear correlation   9
linear dynamic range   5
liver   13
LNA™-enhanced arrays   26
LNCaP cells   125
locked nucleic acid ISH   119
locked nucleic acids   26
loess normalization   38
luciferase assay   102
lung cancer   199
lymphoid differentiation   96

massively parallel sequencing   83
mathematical models   219
maturation of microRNAs   55
melting temperature (Tm)   58
MetaCore   191
Michaelis-Menten kinetics   223

microarray   2
microarray probe   3
miR-146a polymorphism   91
miR-17-92 cluster   103
miRAGE   86
miRNA   2
miRNA binding to target mRNA   233
miRNA expression   2
miRNA expression signature   190
miRNA inhibitor and hairpin
    libraries   131
miRNA polymorphisms   83
miRNA probe design   27
miRNA processing enzymes   129
miRNA seed region   170
miRNA target prediction models   171
miRNA transfection experiment   230
miRNA-mediated network   175
miRNA-mediated network motifs   231
miRNA-seq   89
miRNeasy kit   14
miRome architecture   153
miRtrons   86
mirWIP1   72
multiplicity of a miRNA-binding
    sites   226
myeloid progenitors   96

NCI-60 cell panel   33
NetLogo   145
network motif   176
network hubs   175
non-cooperative regulation   224
non-steady-state behavior   227
normalization schemes   16, 37

Oblimersen   209
ordinary differential equations   222
OR-gate   177

P38 MAPK pathway   207
pancreatic cancer   199

pathway enrichment analysis 196
P-bodies 221
PicTar 172
PITA 172
Placenta 13
post-transcriptional operon 142
post-translational modification 205
power analysis 59
primer design 57
pro-apoptotic genes 127
promoter occlusion 139
prostate cancer (CaP) 111
prostate cancer-associated protein6 140
prostate specific antigen (PSA) 112
protein-protein interaction network 181
proto-oncogenes 152
PTEN protein 103
pyrosequencing technology 86

quantile normalization 40

RAS oncogenes 102
real-time QPCR array 54
real-time quantitative PCR 3, 53
Rho-activated protein kinase 126
RNA 2
RNA isolation protocol 14
RNAhybrid 172
RNA-seq 91

sample size calculations 59
Sanger miRBase 116

scale free topology 175
SCIO 209
SILAC experiments 126
Single Input Motif (SIM) 230
Solexa sequencing 87
spiked-in DNA or RNA 17
statistical noise 2
stem-progenitor cells 96
syntenic chromosomal regions 153
systems biology 189

T-cell acute lymphoblastic lymphoma 156
TarBase 191
target validation 126
targets of miRNAs 19
TargetScan 172
T-cell development 96
TF binding motifs 174
Transfac 174
TF-miRNA feedback circuit 232
threshold-linear response 220
total RNA 3
transcript degradation rate 223
transcription factors 169
Transfac 174
two-class model of gene regulation 220
two-color microarray 37

un-templated 3' nucleotides 84

validated microRNA/target interactions 104
viral integration sites 152